Small Engines:
Operation
and Maintenance

Other books by William H. Crouse

Automotive Chassis and Body
Automotive Electrical Equipment
Automotive Emission Control
Automotive Engine Design
Automotive Engines
Automotive Fuel, Lubricating, and Cooling Systems
Automotive Mechanics
Automotive Mechanics Test Book
Automotive Service Business: Operation and Management
Automotive Transmissions and Power Trains
Workbook for Automotive Chassis
Workbook for Automotive Electricity
Workbook for Automotive Engines
Workbook for Automotive Service and Trouble Diagnosis
Workbook for Automotive Tools
Workbook for Automotive Transmissions and Power Trains
Workbook for Small Engines
General Power Mechanics (*with Robert M. Worthington*
and Morton Margules)
Science Marvels of Tomorrow
Small Engines: Operation and Maintenance
Understanding Science

Automotive Transparencies by William H. Crouse and Jay D. Helsel

Automotive Brakes
Automotive Electrical Systems
Automotive Engine Systems
Automotive Transmissions and Power Trains
Automotive Steering Systems
Automotive Suspension Systems
Engines and Fuel Systems

Small Engines: Operation and Maintenance

William H. Crouse

McGRAW-HILL BOOK COMPANY

NEW YORK	KUALA LUMPUR	PANAMA
ST. LOUIS	LONDON	RIO DE JANEIRO
SAN FRANCISCO	MEXICO	SINGAPORE
DÜSSELDORF	MONTREAL	SYDNEY
JOHANNESBURG	NEW DELHI	TORONTO

ABOUT THE AUTHOR

Behind William H. Crouse's clear technical writing is a background of sound mechanical engineering training as well as a variety of practical industrial experience. He spent a year after finishing high school working in a tinplate mill; summers, while still in school, working in General Motors plants; and three years working in the Delco-Remy Division shops. Later he became Director of Field Education in the Delco-Remy Division of General Motors Corporation, which gave him an opportunity to develop and use his natural writing talent in the preparation of service bulletins and educational literature.

During World War II, he wrote a number of technical manuals for the Armed Forces. After the war, he became Editor of Technical Education Books for the McGraw-Hill Book Company. He has contributed numerous articles to automotive and engineering magazines and has written several outstanding books: "Automotive Mechanics," "Automotive Engine Design," "Electrical Appliance Servicing," "Everyday Automobile Repairs," "Everyday Household Appliance Repairs," "Science Marvels of Tomorrow," and "Understanding Science." He served as the Editor-in-Chief on the first edition of the 15-volume "McGraw-Hill Encyclopedia of Science and Technology." He is a coauthor of "General Power Mechanics." He has also contributed eight other textbooks to the McGraw-Hill Automotive Technology Series.

William H. Crouse's outstanding work in the automotive field has earned for him membership in the Society of Automotive Engineers and in the American Society for Engineering Education.

Library of Congress Cataloging in Publication Data
Crouse, William Harry, 1907–
  Small Engines.

  1. Gas and oil engines. I. Title.
TJ785.C76   621.43'4   73-2631
ISBN 0-07-014691-8

567890 DODO 798

The editors for this book were Ardelle Cleverdon and
Cynthia Newby, the designer was Marsha Cohen,
and its production was supervised by James E. Lee.
It was set in Century by Typographic Sales, Inc.
The printer and binder was R. R. Donnelley & Sons Company.

Contents

Preface

It seems that almost every day engineers, technicians, mechanics, and just plain handymen find new uses for small engines. Everyone knows how the small, single-cylinder engine has revolutionized mowing the lawn, cutting down trees and sawing them up, removing snow from the walk and driveway, and tilling the garden. In addition to power mowers, chain saws, snow blowers, and soil tillers, the small engine will be found doing a hundred other jobs around the home and farm. And it has opened up new horizons to the sports-minded with, for example, the snowmobile, the outboard motor, and the miniature race car. The list could go on and on.

More and more people are buying engine-powered equipment because of its ease of operation, its simplifying of tedious chores around the house and farm, the added pleasure the equipment gives to their chosen sport. And even the person with the "postage-stamp" yard is apt to have a power mower and edger because they do such a neat job with such little effort.

All this means that millions of small engines must be built each year. Experts estimate that more than 70 million small engines are in use in the United States today, and that about 6 million more come off the assembly line each year. It is also true that millions of small engines are junked each year. Some are worn out, some die from lack of service and abuse. Many of these engines die prematurely; a little service at the right time could prolong their lives.

This is one purpose of the book you have in your hands—to give you the background information you need to service any and all small engines. The second purpose of the book is to give you a profitable skill so you can improve your position in life. The small-engine expert, who can handle any kind of small-engine service problem, is in demand, and as you study this book, you are fitting yourself for a profitable career as a small-engine technician.

The engineers who have designed small engines have been very ingenious. They have found several ways to crank the small engine. They have come up with such innovations as a combined magneto and alternator which not only furnishes the sparks to keep the engine running, but also furnishes the electric current to keep the battery charged. They have developed numerous special and simple carburetors for the variety of fuel systems that small engines require. You will find other interesting and unique mechanisms described as you study this book.

The information in this book, combined with practical experience in the shop, will fit you for a career as a small-engine technician. To assist you in your shop training, a "Workbook for Small Engines" has been developed. This workbook is correlated with the "Small Engines" text. Together, the two books provide a complete course in the operation and servicing of small engines.

The author wishes to thank the American Association for Vocational Instructional Materials as publishers of "Small Engines, Vol. 1: Care and Operation, and Vol. 2: Maintenance and Repair," from which many illustrations have been borrowed or adapted for use in this publication.

William H. Crouse

How to Study This Book

This book is designed to give you the complete background information you need to become a small-engine technician. Here are several things you can do to help you understand and remember what you are studying.

GETTING PRACTICAL EXPERIENCE

This book alone will not make you an expert in small engines, just as books alone cannot make an airplane pilot or an architect or a dentist the expert he is. You also need practice, practice in handling engine parts and the tools of the trade. If you are taking a regular course in small-engine servicing, you will get this practice under the supervision of an expert teacher. But if you are not taking a regular course, you can still get practical experience in a local engine-repair shop where small engines are being serviced. If you are already working in a repair shop, then this book will broaden your knowledge and fit you for greater responsibilities.

SERVICE PUBLICATIONS

While you are in the repair shop, try to get a chance to study the various service publications it receives on small-engine repair and maintenance work. All small-engine manufacturers publish service manuals on their engines which are designed to help the small-engine repairman do a good service and repair job on their engines. Studying these publications will be very helpful to you.

KEEPING A NOTEBOOK

You will find that keeping a notebook will be of great value to you. Get a loose-leaf notebook so you can insert or remove pages and thereby add to and improve your notebook. Write the answers to the questions you will find in the tests at the ends of the chapters in the book. Write down any interesting procedure you run across in your shop work, or in studying the book or manufacturers' service manuals. As you do all this, you will find that your book will become an increasingly valuable source of information you can refer to.

DOING THE JOB

Keep your eyes open, remember what you see and hear about the operation and servicing of small engines. Write down the important facts in your notebook. Learn to do the various servicing jobs quickly and properly. All this is important. In addition, be pleasant and cooperative with your fellow workers, your superiors, the customers you meet at your place of business. To really get ahead in this world, you must not only know how to do your job well, but you must also know how to get along with people and make them like you. It is the fellow who knows his stuff and is well-thought of by others who gets ahead in this world. Let this fellow be you!

William H. Crouse

Acknowledgments

During the years that this book was in the planning stage, and later during its actual writing and illustrating, the author was given invaluable aid and inspiration by many, many people in the small-engine field and in education. The author gratefully acknowledges his indebtedness and offers his sincere thanks to these many people. All cooperated with the aim of providing accurate and complete information on how small engines are constructed, how they operate, and how to maintain and service them. Special thanks are due to the following organizations for information and illustrations they supplied: AC Spark Plug Division, Chevrolet Motor Division, Delco-Remy Division, and Pontiac Division, all of General Motors Corporation; American Association for Vocational Instructional Material; Briggs and Stratton Corporation; Chrysler Motors Corporation; Clinton Engines Corporation; Cushman Motors; Federal-Mogul-Bower Bearings, Inc.; Ford Motor Company; Gravely Tractors; Homelite Division of Textron, Inc.; International Harvester Company; Jacobsen Manufacturing Company; Johnson Motors; Kohler Company; Lawn Boy Division of Outboard Marine Corporation; Lawson; McCulloch Corporation; Merry Manufacturing Company; Onah Division of Studebaker Corporation; Power Curbers Inc.; Tecumseh Products Company; Tennant Company; Lauson-Power Products Engine Division; Volkswagen; West Bend Company; Wisconsin Engine Company.

To all these organizations and the people who represent them, sincere thanks!

William H. Crouse

Small-Engine Applications

This is a book about small engines, the type of engine used in lawn mowers, edgers, minibikes, snow blowers, chain saws, and numerous other devices such as post-hole diggers, irrigation pumps, air compressors, water pumps, sprayers, grinders, and so on. The variety of jobs that small internal-combustion engines do is amazing and nearly endless. It is estimated that there are more than 70 million small engines now in use in the United States, and at least 6 million more come off the assembly line each year. Several million small engines are junked each year. Some really wear out. Many could give additional service if they were serviced and repaired. And many would last much longer if the owners did not abuse them and gave them decent care. This is what the book is all about—how to take care of small engines properly so they will give the life that is built into them. First, we describe various types of small engines, explain the different machines in which they are used. Then we will discuss the construction and operation of small engines, and describe the various adjustments, services, and repairs that can be made on them.

Figures 1.1 to 1.18 show various small engines and some of the equipment they operate. As you will note from studying these illustrations, small engines have been adapted to all sorts of mechanical equipment. And the illustrations shown in this chapter represent only a small part of the total picture of small-engine use. You could fill a book with just pictures showing all the various uses of small engines.

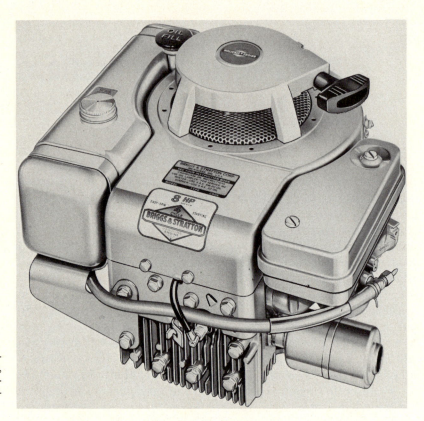

FIGURE 1.1 An 8-horse-power, single-cylinder, four-cycle, air-cooled engine. (*Briggs and Stratton Corporation.*)

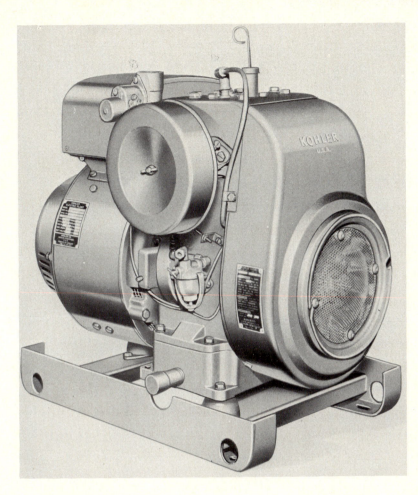

FIGURE 1.2 Electric power generator operated by a single-cylinder engine. (*Kohler Company.*)

FIGURE 1.3   Portable
water pump operated by
a small, one-cylinder en-
gine. (*Clinton Engine Corporation.*)

FIGURE 1.4 Power mower operated by a small engine. (*Jacobsen Manufacturing Company.*)

FIGURE 1.5   Riding mower powered by a small engine.   (*International Harvester Company.*)

FIGURE 1.6 Snow thrower, powered by a small engine, shown in action. (*Jacobsen Manufacturing Company.*)

FIGURE 1.7 Rotary soil tiller. The engine drives the rotating blades which till the soil. (*Merry Manufacturing Company.*)

FIGURE 1.8 Sickle mower. (*Merry Manufacturing Company.*)

FIGURE 1.9 Minibike. The one-cylinder engine, driving the rear wheel through a belt, gives the minibike respectable performance. (*Clinton Engines Corporation.*)

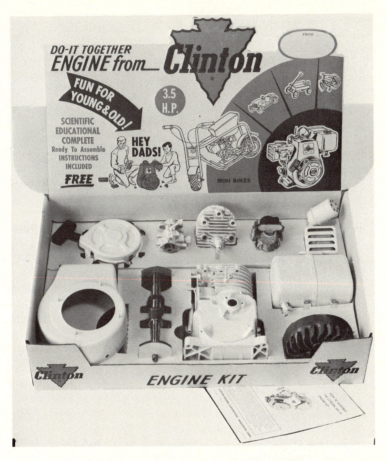

FIGURE 1.10 You can buy an engine kit from which you can build an actual 3.5-horsepower engine. (*Clinton Engines Corporation.*)

FIGURE 1.11 A small engine can take you out to the golf course. The engine, heavily muffled, makes very little noise that would disturb the golfers. (*Clinton Engine Corporation.*)

FIGURE 1.12 The engine in a chain saw is a special adaptation of the basic small-engine design. It can be used in any position. (*Homelite Division of Textron Inc.*)

FIGURE 1.13 A partial cutaway view of a chain saw, showing the internal construction of the engine. (*Homelite Division of Textron Inc.*)

FIGURE 1.14 A small engine powers this one-man police vehicle. (*Cushman Motors Division of Outboard Marine Corporation.*)

FIGURE 1.15   An electric welder powered by a small engine. (*Lincoln Electric Company.*)

FIGURE 1.16   A small engine operates this power curber, which forms a curb, as shown, when the hopper is filled with concrete and the machine is moved along in the required direction. (*Power Curbers, Inc.*)

FIGURE 1.17 Powered floor sweeper which uses a single-cylinder engine. (*Tennant Company.*)

FIGURE 1.18   Power scrubber. This equipment uses a two-cylinder engine to operate it. It carries a large tank filled with cleaning solution which is sprayed on the floor. A pair of rotating brushes scrubs the floor clean and a vacuum intake pulls in the debris and deposits it in a hopper. (*Tennant Company.*)

Engine Principles | 2

In this chapter we will look at the fundamental principles that make the engine "go." These basic principles are not complicated. They are easy to understand. Once you understand them, you will have no trouble understanding how engines work. And this is important because when you know how and why engines work, you will find it easier to find trouble causes when something goes wrong. Knowing what is causing a trouble makes it much easier to fix the trouble. Thus, the material in this chapter is important to you. It is one key to fast, efficient, and profitable small-engine service work.

2.1 Atoms. You might think it strange for us to start this chapter, on engine principles, with an explanation of atoms. But an engine will not run until atoms start getting together inside the engine. Therefore, we should take a close look at atoms.

You can make a list a mile long of all the different things you see around you—this book, your chair, the window, the trees or buildings outside, the clouds, and so on. And all these things are made of metals, wood, paper, glass, cloth, leather, clay, water, air, and thousands of other materials. But amazingly enough, all these different things are made of only a few basic "building blocks" called atoms. Of course, atoms are not really "blocks" as we will learn when we study them. There are only about 100 different kinds of atoms. But these 100 kinds of atoms can be put together in millions of different ways to form millions of different substances. You can compare this with the 26 letters of our alphabet. These letters can be put together in many different ways to make up the several hundred thousand words in our language.

Now, about those 100 or so kinds of atoms. We have special names for each kind, such as copper, iron, carbon, oxygen, silver, gold, uranium, aluminum, and mercury. The silver in the dime in your pocket is made up of an almost countless number of one kind of atom. The oxygen in the air you breathe and in the water you drink is made up of a vast number of another kind of atom. Any substance made up of only one kind of atom is called an element. Silver is an element. So are oxygen, and hydrogen, and sodium, and all the others listed in the table of elements (Fig. 2.1). Actually, the table lists only a few of the more common elements.

TABLE OF ELEMENTS

Name	Symbol	Atomic number	Approximate atomic weight	Electron arrangement
Aluminum	Al	13	27	·2)8)3
Calcium	Ca	20	40	·2)8)8)2
Carbon	C	6	12	·2)4
Chlorine	Cl	17	35.5	·2)8)7
Copper	Cu	29	63.6	·2)8)18)1
Hydrogen	H	1	1	·1
Iron	Fe	26	56	·2)8)14)2
Magnesium	Mg	12	24	·2)8)2
Mercury	Hg	80	200	·2)8)18)32)18)2
Nitrogen	N	7	14	·2)5
Oxygen	O	8	16	·2)6
Phosphorus	P	15	31	·2)8)5
Potassium	K	19	39	·2)8)8)1
Silver	Ag	47	108	·2)8)18)18)1
Sodium	Na	11	23	·2)8)1
Sulfur	S	16	32	·2)8)6
Zinc	Zn	30	65	·2)8)18)2

FIGURE 2.1 Table of some of the more common elements.

2.2 Size of Atoms. Atoms are very small. In a single drop of water there are more than 100 billion billion atoms. This is about 30 billion atoms for every person living on the earth. If you tried to count your share — your 30 billion atoms — it would take you 1,000 years if you counted one atom every second, day and night. And this is only your share of just one drop of water.

2.3 Inside the Atom. Now let us, in our imagination, look inside atoms to see what they are made of. You are likely to be disappointed. For there is almost nothing inside the atoms. Take, for example, the hydrogen atom. It is made up of only two particles. One of these is at the center, or nucleus, of the atom. The other, a comparatively long distance away, is whirling in an orbit around the nucleus. The center particle is called a proton. The outside particle, in orbit around the proton, is called an electron.

Suppose that the proton were the size of a marble. If you laid this marble under the basket at one end of a standard high school basketball court (84 ft long), the electron would be as far away as the other basket. And there would be nothing in between. It is all just empty space.

The proton has a tiny charge of positive electricity [indicated by a plus (+) sign]. The electron has a tiny charge of negative electricity [indicated by a minus (−) sign]. Opposites attract. Minus attracts plus. Plus attracts minus. Thus, the negatively charged electron is pulled toward the positively charged proton. But balancing this inward-pulling force is the outward pull of centrifugal force. This is somewhat like the balancing of forces you get when you whirl a ball on a rubber band in a circle around your hand (Fig. 2.2). The rubber

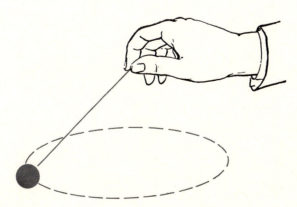

FIGURE 2.2 The electron in a hydrogen atom circles the proton like a ball on a rubber band swung in a circle around the hand.

band pulls the ball toward your hand. But the centrifugal force pushes the ball away. The result is that the ball moves in an orbit, or in a circle, around your hand.

2.4 Helium.   The simplest atom is hydrogen. It has one proton and one electron. Next, as we go from the simplest to the more complex atoms, is helium, another gas. The helium atom has two protons (+ charges) in its nucleus and two electrons (− charges) circling the nucleus (Fig. 2.3). In addition, the nucleus has two other particles which are electrically neutral (have no charge) and are therefore called *neutrons*. The neutrons weigh almost the same as the protons. They seem to serve as a sort of nuclear "glue" to keep the two protons together in the nucleus. Like electrical charges repel each other, and, without the neutrons, the protons would fly apart. But the presence of the neutrons in the nucleus seems to nullify this repulsive force between the protons so that they stay together inside the nucleus.

2.5 More Complex Atoms.   The next element after helium in complexity is lithium, a very light metal. The lithium atom (Fig. 2.4) has a nucleus with three protons and four neutrons. Three electrons, one for each proton, circle the nucleus.

Next is beryllium, another metal, with four protons, four neutrons, and four electrons; boron with five protons, five neutrons, and five electrons; carbon with six, six, and six; nitrogen with seven, seven, and seven; oxygen with eight, eight, and eight, and so on. Note that each atom normally has the same number of electrons as protons. This makes the atom electrically neutral, since negative charges equal

FIGURE 2.3   Helium atom has two electrons, two protons, and two neutrons.

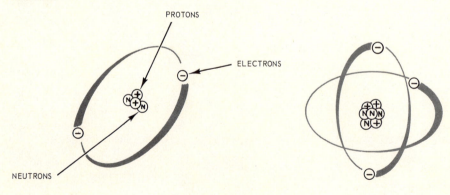

PROTONS

ELECTRONS

NEUTRONS

positive charges. However, as we will soon learn, some kinds of atoms are not always able to hold on to all their electrons. These electrons therefore "wander" off, leaving electrically unbalanced atoms behind (with + charges). The ability of electrons to free themselves from atoms in this manner gives us the phenomenon of electricity, as we will learn later.

2.6 Chemical Reactions. When two or more atoms link up, or combine, they form a molecule. The linking-up process is called chemical reaction. For example, two atoms of hydrogen and one atom of oxygen react to form one molecule of water (Fig. 2.5). Water has the chemical formula H_2O, which means each molecule has two atoms of hydrogen and one atom of oxygen. When one atom of sodium (chemical symbol Na) unites with one atom of sodium (chemical symbol Cl), a molecule of common table salt is formed (NaCl). Another example is sugar, each molecule of which has 12 atoms of carbon, 22 atoms of hydrogen, and 11 atoms of oxygen and the chemical formula $C_{12}H_{22}O_{11}$.

During a chemical reaction, one or more of the electrons in the outer shells of some of the atoms are shared with other atoms. This matter of sharing will become clearer as we discuss electron shells, or orbits.

2.7 Combustion. Combustion, or fire, is a common chemical reaction in which the gas oxygen combines with other elements, such as hydrogen or carbon. One type of combustion occurs in the automobile engine. A mixture of air and gasoline vapor is compressed

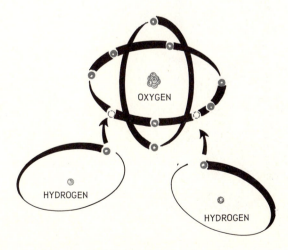

FIGURE 2.5 One atom of oxygen uniting with two atoms of hydrogen to form a molecule of water, or H_2O.

and then ignited, or set on fire. The air is about 20 percent oxygen. Gasoline is mostly hydrogen and carbon (and is thus called a hydrocarbon). The chemical reaction during combustion is between the three elements of oxygen, hydrogen, and carbon.

We have seen how one oxygen atom combines with two hydrogen atoms to form H_2O, or water. Similarly, one carbon atom combines with two oxygen atoms to form a molecule of CO_2, or carbon dioxide (Fig. 2.6). Carbon dioxide is a gas. The carbon atom has six electrons in two shells, two electrons in the inner shell and four in the outer shell. When an atom of carbon combines with two atoms of oxygen, each oxygen atom takes two of the carbon atom outer-shell electrons.

Thus, during the combustion process, oxygen in the air combines with the carbon and hydrogen in the gasoline to form carbon dioxide and water vapor. During the process, temperatures go as high as

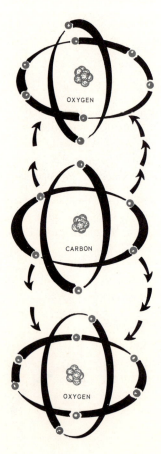

FIGURE 2.6 Two atoms of oxygen uniting with one atom of carbon to form a molecule of carbon dioxide, or CO_2.

6000° F. Both the water, in vapor form, and the carbon dioxide, a gas, leave the engine through the exhaust system.

NOTE: With ideal combustion, all the hydrogen and carbon are converted to H_2O and CO_2. In the engine, however, ideal combustion is not achieved, and some hydrocarbons are left over. Also, some carbon monoxide (CO) is produced instead of CO_2. These products contribute to the smog problem.

2.8 Heat.   We mentioned that combustion is accompanied by high temperatures. But what do we mean by "high temperature" or "heat"? From the scientific point of view, heat is simply the rapid motion of atoms or molecules in a substance.

This is not easy to understand. First, we must understand that the atoms and molecules of any substance are in rapid motion. Even though a piece of iron appears solid and motionless, the atoms and molecules in the iron are in rapid motion. The atoms in a piece of hot iron are moving faster than those in a piece of cold iron.

2.9 Change of State.   Maybe we can clear up the matter by discussing change of state. If we put a pan of ice cubes over a fire, the ice cubes soon melt, or turn to water. Then, the water boils, or turns to vapor. Most substances can exist in any of three states — solid, liquid, or gas (or vapor). When a substance changes from one state to another, it undergoes a change of state.

A change in the speed of molecular motion, if great enough, results in a change of state. For example, in ice, the water molecules are moving slowly and in restricted paths. But as the temperature increases, the molecules move faster and faster. Soon, the molecules are moving so fast that they break out of their restricted paths. The ice turns to water at 32° F. As molecular speed increases still more, the boiling point is reached (212° F at sea level). Now the molecules are moving so fast that great numbers of them fly clear out of the water. The water boils, or turns to vapor.

2.10 Producing Change of State.   But what makes the water molecules move faster? Here is a simple explanation. During combustion, oxygen unites with carbon or hydrogen atoms. The new molecules thus formed are set into extremely rapid motion. The rushing together of the atoms to satisfy the unbalance of electric charges can be said to produce this rapid motion. Now, the newly formed and rapidly moving molecules from the fire below the pan bombard the pan. This bombardment sets the molecules of metal in the pan into rapid motion (the pan becomes hot). The metal molecules, in turn, bombard the ice molecules. The ice melts and then turns to vapor.

2.11 Light and Heat Radiations. This is only a partial description of what takes place during combustion. In addition to the swiftly moving molecules, the fire also produces radiations. We can see these radiations as light and feel them as heat. They are produced by certain actions inside the atoms of fuel and oxygen. A partial explanation of these actions is that the inner electrons of the atoms are disturbed by the actions of the outer electrons as they move between atoms. The inner electrons jump between shells. Each jump is accompanied by a tiny flash of radiant energy.

2.12 Expansion of Solids with Heat. When a piece of iron is heated, it expands. A steel rod that measures 10 ft in length at 100° F will measure 10.07 ft in length at 1000° F (Fig. 2.7). Here is the reason. As the rod is heated, the molecules in it move faster and faster. They need more room for this and therefore push adjacent molecules away so that the rod gets longer.

2.13 Expansion of Liquids and Gases. Liquids and gases also expand when heated. A cubic foot of water at 39° F will become, when heated to 100° F, 1.01 cubic feet. A cubic foot of air at 32° F, heated to 100° F without a change of pressure, will become 1.14 cubic feet. These expansion effects result from more rapid molecular motion, which tends to push the molecules farther apart so that they spread out and take up more room.

2.14 Increase of Pressure. A different sort of effect results if the volume is held constant while the cubic foot of air is heated from 32 to 100° F. If we start with a pressure of 15 psi (pounds per square inch), we find that the pressure increases to about 17 psi at 100° F. This can also be explained by the molecular theory of heat.

Gas pressure in a container is due entirely to the unending bombardment of the gas molecules against the inside of the container (Fig. 2.8). As we have already said, gas molecules move about in all directions at high speeds. They are continually bumping into one

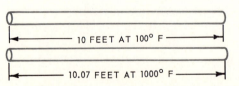

FIGURE 2.7   Steel rod that measures 10 ft. at 100° F will measure 10.07 ft. at 1000° F.

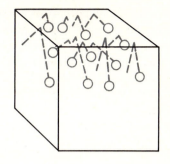

FIGURE 2.8 Gas pressure in a container is the result of the ceaseless bombardment of the inner sides of the container by the fast-moving molecules of gas. This bombardment is shown on only one side of the container for simplicity. It actually takes place against all the inner sides. The molecules are shown tremendously enlarged. Also, of course, there are billions of molecules entering into the action, not just a few, as shown.

another and into any solid that is in their way. Thus, the walls of the container are bumped by these countless billions of molecules. These "bumps" add up to a combined push, or pressure.

As temperature increases, the molecules of gas move faster. They bump the walls of the container harder and more often. The result is higher pressure in the container.

Another way to increase pressure in a container is to compress the gas in the container into a smaller volume. This is what happens in engine cylinders. The mixture of air and gasoline vapor is squeezed to about one-ninth or one-tenth of its original volume. The molecules are moving much faster, hitting the cylinder head and piston more often and faster. The pressure goes up.

Still greater pressure is then achieved in the engine cylinder by igniting the compressed air-fuel mixture. When this happens, the mixture burns, as has already been explained, and the temperature of the burning gas goes as high as 6000° F. This means that the gas molecules are moving at very high speed. They hit the top of the piston so hard and so often that a push, or pressure, of a ton or more may be registered on the piston. This pressure, or push, is due entirely to countless fast-moving molecules bombarding the piston.

2.15 Increase of Temperature. Not only pressure but also temperature increases when a gas is compressed. Moving the molecules closer together causes them to bump into one another more often so that they are set into faster motion. Faster motion means a higher temperature. For example, in the diesel engine, air is compressed to as little as one-sixteenth of its original volume. This raises the temperature of the air to as much as 1000° F. Of course, the heat produced by the action soon escapes from the compressed air and its container into the surrounding air. Any hot object loses heat until its temperature falls to that of the surrounding medium.

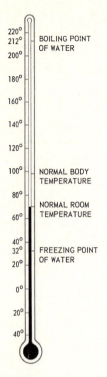

FIGURE 2.9 Fahrenheit thermometer.

2.16 The Thermometer. The thermometer (Fig. 2.9) shows a familiar use of the expansion of liquids as temperature goes up. The liquid, usually mercury (a metal that is liquid at ordinary temperatures), is largely contained in the glass bulb at the bottom of the glass tube. As temperature increases, the mercury expands. Part of it is forced up through the hollow glass tube. The higher the temperature, the more the mercury expands and the higher it is forced up through the tube. The tube is marked off to indicate the temperature in degrees.

2.17 The Thermostat. Different metals expand at different rates with increasing temperatures. Aluminum expands about twice as much as iron as their temperatures go up. This difference in expansion rates is used in thermostats. Thermostats do numerous jobs in small engines and in automobiles. One type is shown in Fig. 2.10. It consists of a coil made up of two strips of different metals, brass and steel, for example, welded together. When the coil is heated, one metal expands faster than the other, causing the coil to wind up or unwind.

2.18 Gravity. Gravity is the attractive force between all objects. When we release a stone from our hand, it falls to earth. When a car is driven up a hill, part of the engine power is being used to lift the car against gravity. Likewise, a car can coast down a hill with the engine turned off, because gravity pulls downward on the car.

Gravitational attraction is usually measured in terms of weight. We put an object on a scale and note that it weighs 10 lb. What we mean is that the object has sufficient mass for the earth to register this much pull on it. Gravitational attraction gives any object its weight.

FIGURE 2.10 Coil-type thermostat. The coil winds up or unwinds as the temperature goes up or down. The resulting motion can be used to operate a control.

2.19 Atmospheric Pressure. The air is also an "object" that is pulled toward the earth by gravity. At sea level and average temperature, a cubic foot of air weighs about 0.08 lb, or about 1.25 oz. This seems like very little. But the blanket of air—our atmosphere—surrounding the earth is many miles thick. This means that there are, in effect, many thousands of cubic feet of air piled on top of one another, all adding their weight. The total weight, or downward push, of this air amounts to about 15 psi at sea level. The pressure of all this air pushing downward is about 2,160 pounds on every square foot. Since the human body has a surface area of several square feet, it has a total pressure of several tons on it.

It would seem that this tremendous pressure would crush you. The reason that it does not is that the internal pressures inside the body balance the outside pressure. Fish have been found thousands of feet below the surface of the ocean, where pressures are more than 100,000 psi (or 700 tons per square foot). The fish can live because their internal pressures balance these immense outside pressures.

2.20 Vacuum.   A vacuum is the absence of air or any other matter. Astronauts, on their way to the moon and the other planets, soon leave the blanket of air surrounding the earth and pass into the vast region of empty space. Out in space, there are only a few scattered atoms of air. This is a vacuum.

FIGURE 2.11   Barometer. The mercury in the tube will stand at about 30 in. above the surface of the mercury in the dish at an atmospheric pressure of 15 psi.

2.21 Producing a Vacuum.   There are many ways to produce a vacuum. The engine, as it operates, produces a partial vacuum in the engine cylinder. The fuel pump works by producing a partial vacuum.

1. Barometer.   The mercury barometer is another device that utilizes a vacuum. You can make a barometer by filling a long tube with mercury and then closing the end. Next, turn the tube upside down, and put the end in a dish of mercury. Now, open the end. Some of the mercury will run down out of the tube, leaving the upper end of the tube empty (a vacuum). (See Fig. 2.11.)

The barometer is used to measure atmospheric pressure. When atmospheric pressure increases, the increased push on the mercury forces it higher in the tube (Fig. 2.12). When atmospheric pressure goes down, the mercury also goes down in the tube. The barometer is used to forecast weather. Before a storm, the atmospheric pressure usually drops. This is due to the heated and therefore lighter air accompanying a storm. Thus, when the mercury falls in the barometer, it indicates that a storm is coming.

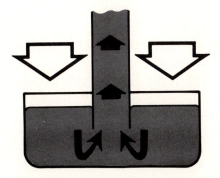

FIGURE 2.12 The pressure of the air, acting on the surface of the mercury and through the mercury, holds the mercury up in the tube. If the air pressure increases, the mercury will be forced higher in the tube.

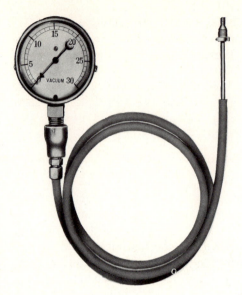

FIGURE 2.13   Vacuum gauge.

2. Vacuum Gauge. The vacuum gauge is really a pressure gauge. The type of vacuum gauge used in engine service contains a bellows or diaphragm which is linked to an indicating needle on the dial face (Fig. 2.13). When the vacuum gauge is connected to the engine (to the intake manifold), the vacuum produced by the engine causes the needle to move and register the amount of vacuum. This action results because the vacuum causes part of the air in the bellows or back of the diaphragm to pass into the engine. Then, air pressure causes the bellows or diaphragm to move, thereby causing the needle to move. The amount that the needle moves depends on the amount of vacuum. The amount of vacuum that a running engine can produce is a measure of engine condition. The vacuum gauge is thus a good diagnostic tool to determine the actual condition of the engine. This is discussed in detail in a later chapter.

CHECKUP

The general engine principles you have been studying in the chapter you just completed are important because they will help you to understand why engines work. Also, they will help you to diagnose engine troubles and fix them because you will know the "why" as well as the "how" of engine operation. The following questions will not only give you a chance to check up on how well you understand and remember these fundamentals, but also will help you to remember them better. The act of writing down the answers to the questions will fix the facts more firmly in your mind.

NOTE: Write down your answers in your notebook. Then later you will find your notebook filled with valuable information which you can refer to quickly.

Completing the Sentences: Test 2.   The sentences below are not complete. After each sentence there are several words or phrases, only one of which will correctly complete the sentence. Write each sentence in your notebook, selecting the proper word or phrase to complete it correctly.

1. The smallest particle into which an element can be divided is called (*a*) a molecule; (*b*) an electron; (*c*) an atom.
2. The reason the atmospheric pressure goes down when air temperature goes up is that (*a*) heated air is heavier; (*b*) heated air is lighter; (*c*) cold air is lighter.
3. The electron has a charge of (*a*) positive electricity; (*b*) neutral electricity; (*c*) negative electricity.
4. When you heat an object, you cause its (*a*) molecules to move faster; (*b*) speed to increase; (*c*) molecules to atomize.
5. During combustion, atoms of such elements as hydrogen or carbon unite with atoms of (*a*) gasoline; (*b*) oxygen; (*c*) H_2O.
6. Gasoline is called a hydrocarbon because it is made up essentially of (*a*) carbon and hydrogen; (*b*) carbon and oxygen; (*c*) hydrogen and oxygen.
7. Most substances can exist in any of three states (*a*) solid, gas, or vapor; (*b*) solid, ice, steam; (*c*) solid, liquid, vapor.
8. When gasoline burns, two of the main compounds that are formed are (*a*) oxygen and hydrocarbon; (*b*) water and carbon dioxide; (*c*) water and oxygen.
9. If you heated a closed container of air, you would find that inside the container the (*a*) pressure would increase; (*b*) volume would increase; (*c*) pressure would decrease.
10. When you compress a gas, you will find that the (*a*) temperature increases; (*b*) temperature decreases; (*c*) combustion takes place.

Written Checkup

In the following, you are asked to write down, in your notebook, the answers to the questions asked or to define certain terms. Writing the answers down will help you to remember them.

1. What are molecules?
2. What are the major parts of atoms?
3. What is a chemical reaction?
4. What is combustion?
5. What does the term "change of state" mean?
6. Describe combustion in the engine cylinder.
7. Explain what heat is in terms of molecular motion.
8. Explain gas pressure in terms of molecular action.
9. Why does the gas pressure in a container increase with increasing temperature?
10. What is atmospheric pressure?
11. Explain how a barometer works.
12. What is vacuum?

Two-Cycle Engine Operation

3

In this chapter, we will find out how the two-cycle engine works. Later, we will look at the four-cycle engine. Both of these terms— *two-cycle* and *four-cycle*—will be explained later in the book. First, however, we want to say that both of these engines are *internal-combustion* engines. That is, the combustion, or fire, that makes the engines go takes place inside the engine. This is in contrast to the steam engine; the fire for this engine takes place outside the engine, in a separate boiler. The boiler boils water to produce steam and the steam then enters the steam engine to make it run. The steam engine is therefore called an *external-combustion* engine because the combustion, or fire that makes it run takes place outside the engine.

Now, back to internal-combustion engines—two-cycle and four-cycle. As we mentioned, we will look at the two-cycle engine first because it is simpler in construction, has fewer internal parts to wear, and is thus the most widely used engine for such equipment as power lawn mowers, edgers, tillers, power saws, and so on. There are, however, many small four-cycle engines in operation and, of course, about all automotive engines are of the four-cycle type. We will explain the basic differences between the two later. First, however, let us look at the two-cycle engine.

3.1 The Piston and the Cylinder. Imagine a tin can with one end cut out. Imagine a second tin can slightly smaller in size which will fit snugly into the first can. See Fig. 3.1. Now suppose you pushed the smaller can rapidly up into the larger can, trapping air ahead of it. This air would be pushed into a smaller space than it previously occupied. The air would be compressed. If the air contained a small amount of gasoline vapor, and if an electric spark were applied to this compressed air–fuel mixture, there would be an explosion and the smaller can would be blown out of the larger can. See Fig. 3.1.

This is about what happens in the internal combustion engine except that the smaller can is not blown all the way out. Instead, in the actual engine, there is an arrangement to prevent this. In the

FIGURE 3.1 Three views showing the actions in an engine cylinder. (*a*) The piston is a metal plug that fits snugly into the engine cylinder. (*b*) When the piston is pushed up into the cylinder, air is trapped and compressed. The cylinder is drawn as though it were transparent so that the piston can be seen. (*c*) The increase of pressure as the gasoline vapor and air mixture is ignited pushes the piston out of the cylinder.

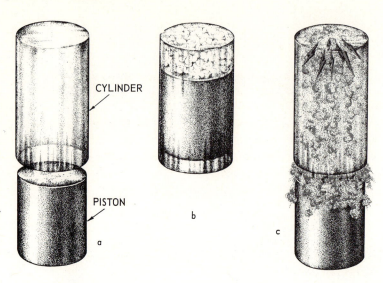

FIGURE 3.2 Typical piston with piston rings in place. When the piston is installed in the cylinder, the rings are compressed into the grooves in the piston.

engine, the larger can is called the cylinder, the smaller can is called the piston. The piston slides up and down in the cylinder.

3.2 Piston Rings. The piston must be a fairly loose fit in the cylinder. If it were a tight fit, it would expand as it got hot and might stick tight in the cylinder. If a piston sticks, it could ruin the engine. On the other hand, if there is too much clearance between the piston and cylinder wall, much of the pressure from the burning gasoline vapor will leak past the piston. This means that the push on the piston will be much less effective. It is the push on the piston that delivers the power from the engine.

To provide a good sealing fit between the piston and cylinder, pistons are equipped with piston rings, as shown in Fig. 3.2. The rings are made of cast iron or other metal. They are split at one point so they can be expanded and slipped over the end of the piston and into the ring grooves which have been cut in the piston. See Fig. 3.3. When the piston is installed in the cylinder, the rings are compressed into the ring grooves so that the split ends come almost together. The rings fit tightly against the cylinder wall and against the sides of the ring grooves in the piston. Thus, they form a good seal between the piston and the cylinder wall. The rings can expand or contract as they heat and cool and still make a good seal. Thus, they are free to slide up and down the cylinder wall. In the two-cycle engine, oil is mixed with the gasoline and this mixture enters the crankcase, as we will explain later. The gasoline goes on up to the combustion chamber where it is burned. Part of the oil covers the cylinder wall so it is kept coated with oil. This allows the rings and pistons to slide up and down the wall easily, with little friction. We will have more to say about oil and friction in a later chapter.

Figure 3.4 shows how the piston ring works to hold in the compression and combustion pressures. The arrows show the pressure from above the piston passing through the clearance between the piston and the cylinder wall. It presses down against the top and against the back of the piston ring, as shown by the arrows. This pushes the piston ring firmly against the cylinder wall and also against the bottom of the piston-ring groove. As a result, there are good seals at both of these points. The higher the pressure in the combustion chamber, the better the seal.

Small two-cycle engines have two rings on the piston. Both are compression rings. Two rings are used to divide up the job of holding the compression and combustion pressures. This produces better sealing with less ring pressure against the cylinder wall.

Four-cycle engines have an extra ring, called the oil-control ring. See Fig. 3.5. As you will learn in the next chapter, four-cycle engines are so constructed that they get much more oil in the cylinder wall than do two-cycle engines. This additional oil must be scraped off to prevent it from getting up into the combustion chamber, where it would burn and cause trouble. More on this later.

FIGURE 3.3 All pistons have grooves cut in them into which the piston rings can be installed.

FIGURE 3.4 Pressure in the combustion chamber, either from compression of the air-fuel mixture or from its combustion, presses the ring against the cylinder wall and the lower side of the piston-ring groove.

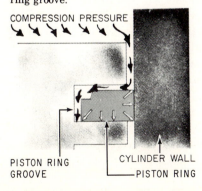

COMPRESSION PRESSURE

PISTON RING GROOVE

CYLINDER WALL

PISTON RING

FIGURE 3.5 The piston of the four-cycle engine has three rings, as shown. The upper two are compression rings, the lower one is an oil-control ring.

OIL SCRAPER RING

COMPRESSION RINGS

OIL CONTROL RING

NOTE: Some four-cycle-engine pistons have four rings; the engine design requires this added ring for adequate oil control.

3.3 The Crank. The piston moves up and down in the cylinder. This up-and-down motion is called *reciprocating* motion. The piston moves in a straight line. This straight-line motion must be changed to rotary, or turning motion, in most machines, before it can do any good. That is, rotary motion is required to make wheels turn, a cutting blade spin, or a pulley rotate. To change the reciprocating motion to rotary motion, a crank and connecting rod are used (Figs. 3.6 and 3.7). The connecting rod connects the piston to the crank.

The crank is a very common device which is used in many machines. For example, the pedal and its support on a bicycle form a crank (Fig. 3.8). The pencil sharpener has a crank (Fig. 3.9). When you put pressure on the foot pedal, or on the sharpener handle, so it swings in a circle, you cause a shaft to turn. In the same way, when the piston is pushed down in the cylinder by the explosion, the push on the piston, carried through the connecting rod to the crank, causes the shaft to turn. Figure 3.10 shows the motions that the piston, connecting rod, and crank go through. As the piston moves up and down, the top end of the connecting rod moves up and down with it. The bottom end of the connecting rod swings in a circle along with the crank.

The piston end of the connecting rod is attached to the piston by a *piston pin,* also called a *wrist pin.* The other end of the connecting rod is attached to the crankpin of the crank by a rod bearing cap. See Figs. 3.6 and 3.7. There are bearings at both ends of the connecting rod so that the rod can move with relative freedom. We will get to bearings later.

NOTE: The crank end of the connecting rod is sometimes called the rod "big end." The piston end of the connecting rod is sometimes called the rod "small end." These are terms you might hear in the repair shop.

3.4 Crankshaft. The crank is part of the crankshaft. It is an offset section, as shown in Fig. 3.11, to which the connecting-rod big end is attached by a bearing. The crankshaft is mounted in the engine on bearings which allow it to rotate. As it rotates, the crank swings in a circle, as shown in Fig. 3.10.

The crankshaft has counterweights, as you will note in Fig. 3.11. These counterweights balance the weights of the crankpin and connecting rod to reduce the tendency of the crankshaft to go out-of-round when it is rotating. This makes for a smoother running engine and much less wear on the bearings which support the crankshaft.

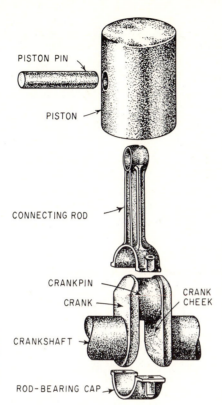

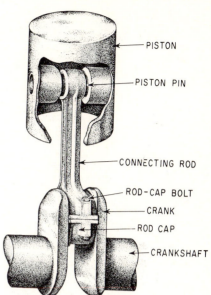

FIGURE 3.6 Piston, connecting rod, piston pin, and crankpin on an engine crankshaft in disassembled view. The piston rings are not shown.

FIGURE 3.7 Piston and connecting-rod assembly attached to the crankpin on a crankshaft. The piston rings are not shown. The piston has been cut away to show how it is attached to the connecting rod.

FIGURE 3.8 The pedals on a bicycle are attached to cranks. The piston, connecting rod, and crank of the engine have been added in shadow to show the comparison between the connecting rod and the lower part of the leg.

FIGURE 3.9 There is a crank on the pencil sharpener.

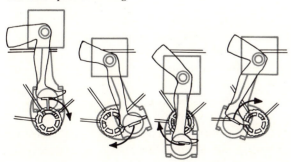

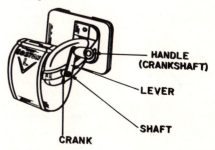

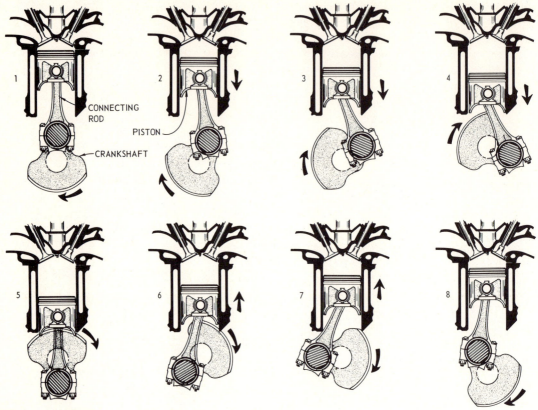

FIGURE 3.10 Sequence of actions as the crankshaft completes one revolution and the piston moves from top to bottom to top again.

3.5 Engine Bearings. The crankshaft is supported by bearings. The connecting-rod big end is attached to the crankpin on the crank of the crankshaft by a bearing. A piston pin at the rod small end is

FIGURE 3.11 The crankshaft converts the reciprocating motion to rotary motion. The crankshaft mounts in bearings which encircle the journals so it can rotate freely.

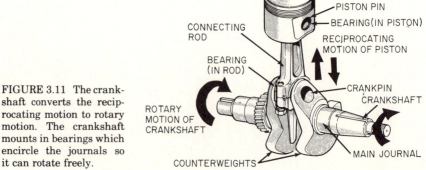

used to attach the rod to the piston. The piston pin rides in bearings. Everywhere there is rotary action in the engine, bearings are used to support the moving parts. The purpose of bearings is to reduce the friction and allow the parts to move easily. Bearings are lubricated with oil to make the relative motion easier. In a later chapter, we will have more to say about friction and engine oil and about the lubricating systems that get the oil to the moving parts.

Bearings used in engines are of two types, sliding or rolling (Fig. 3.12). The sliding type of bearing are sometimes called bushings or sleeve bearings because they are in the shape of a sleeve that fits around the rotating journal, or shaft. The sleeve-type connecting-rod big-end bearings—usually called simply rod bearings—and the crankshaft supporting bearings—called the main bearings—are of the split-sleeve type. They must be split in order to permit their assembly into the engine. In the rod bearing, the upper half of the bearing is installed in the rod, the lower half is installed in the rod-bearing cap. When the rod cap is fastened to the rod, as shown in Fig. 3.7, a complete sleeve bearing is formed. Likewise, the upper halves of the main bearings are assembled in the engine and then the main-bearing caps, with the lower bearing halves, are attached to the engine to complete the sleeve bearings supporting the crankshaft.

The typical bearing half is made of a steel or bronze back to which a lining of relatively soft bearing material is applied. See Fig. 3.13. This relatively soft bearing material, which is made of several materials such as copper, lead, tin, and other metals, has the ability to conform to slight irregularities of the shaft rotating against it. If wear does take place, it is the bearing that wears, and the bearing can be replaced instead of the much more expensive crankshaft or other engine part.

The rolling-type bearing uses balls or rollers between the stationary support and the rotating shaft. See Fig. 3.12. Since the balls

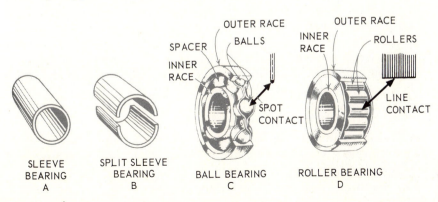

FIGURE 3.12 Sleeve, ball, and roller bearings.

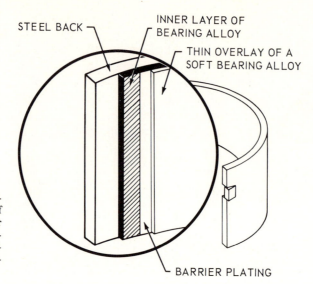

STEEL BACK

INNER LAYER OF
BEARING ALLOY

THIN OVERLAY OF A
SOFT BEARING ALLOY

BARRIER PLATING

FIGURE 3.13 Construction of a bearing half of the sleeve type. The softer bearing material is applied to a hard back. (*Federal-Mogul-Bower Bearings, Inc.*)

or rollers provide rolling contact, the frictional resistance to movement is much less. In some roller bearings, the rollers are so small that they are hardly bigger than needles. These bearings are called needle bearings. Also, some roller bearings have the rollers set at an angle so the races the rollers roll in are tapered. These bearings are called tapered roller bearings (see Fig. 16.42). Some ball and roller bearings are sealed with their lubricant already in place. Such bearings require no other lubrication. Others do require lubrication from the oil in the gasoline (two-cycle engines) or from the engine lubrication system (four-cycle engines).

As you study different small engines, you will find that all the above bearings have been used in different engines. The type of bearing selected by the designers of the engine depends on the design of the engine and the use to which the engine will be put. Generally, sleeve bearings, being less expensive and satisfactory for most engine applications, are used. In fact, sleeve bearings are used almost universally in automotive engines for the main, connecting rod, and piston-pin bearings. But you will find some engines with ball and roller bearings to support the crankshaft and for the connecting rod and piston-pin bearings.

3.6 Making the Engine Run. To make the engine run, we must supply it with a mixture of air and gasoline vapor. This mixture must enter the cylinder and be compressed by the piston as it moves up. Then, we must introduce a spark into the cylinder so the mixture will be ignited. The mixture burns, or explodes, and pushes the piston down. This push is carried through the connecting rod, as we have seen, and this causes the crankshaft to turn. Next, the burned gases must be removed from the cylinder and a fresh charge brought in. These actions continue as long as the engine runs. We will go into detail on these actions later.

3.7 The Piston Stroke. In any piston engine, the movement of the piston from one limiting position to the other is called a *piston stroke.* The upper limiting position of the piston is called *top dead center* (TDC), and the lower limiting position is called bottom dead center (BDC). Thus, a piston stroke takes place when the piston moves from TDC to BDC or from BDC to TDC (Fig. 3.14). When the piston moves from TDC to BDC, after combustion has taken place, the stroke is the *power stroke.* The high pressure of the explosion, forcing the piston to move during the power stroke, results in power from the engine.

3.8 How the Two-Cycle Engine Got Its Name. The full name of the two-cycle engine is two-stroke-cycle engine. The reason for this

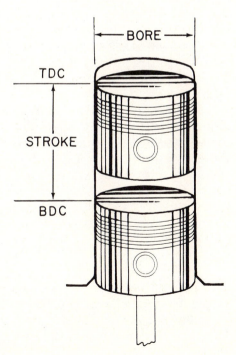

FIGURE 3.14 The bore and stroke of an engine cylinder.

is that it takes two piston strokes, an up stroke and a down stroke, to complete a cycle of engine operation. In other words, everything that happens in the engine takes place in these two strokes and these events continue to be repeated as the engine runs. This is the meaning of *cycle*. A cycle is simply a series of events that repeat themselves. For instance, the cycle of the seasons, spring, summer, fall, and winter, is repeated every year. In a similar way, the two piston strokes in the two-stroke-cycle engine form a cycle that is repeated continuously as the engine runs.

As a rule, the word *stroke* is not used in the name of the engine, so that *two-stroke-cycle* engine has become, in common usage, *two-cycle engine*. Be very careful to remember, however, that there are two piston strokes in a single cycle of the two-cycle engine.

3.9 Engine Actions. The air-fuel mixture enters the cylinder, and burned gases leave the cylinder, through ports, or openings, in the cylinder wall. The port through which air-fuel mixture enters is called the *intake port.* The port through which the burned gases leave is

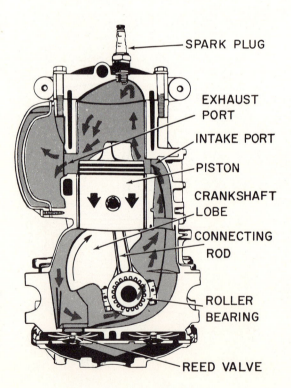

FIGURE 3.15 When the piston moves down past the two ports, air-fuel mixture can flow into the cylinder and burned gases can flow out. (*Johnson Motors.*)

SPARK PLUG

EXHAUST PORT

INTAKE PORT

PISTON

CRANKSHAFT LOBE

CONNECTING ROD

ROLLER BEARING

REED VALVE

called the *exhaust port*. Figure 3.15 shows how this works. When the piston moves down on the power stroke, it clears, or moves past, the ports. Now a fresh air-fuel charge can flow into the cylinder through the intake port. At the same time as the fresh charge is flowing in, burned gases can flow out from the cylinder through the exhaust port.

Let us follow the complete set of actions from the time that the fresh charge of air-fuel mixture is compressed and ignited.

As the piston nears TDC, ignition takes place (Fig. 3.16). The high combustion pressures drive the piston down, and the thrust through the connecting rod against the crank turns the crankshaft. As the piston nears BDC, it passes the intake and exhaust ports in the cylinder wall (Fig. 3.17). Burned gases, still under some pressure,

FIGURE 3.16 Sectional view of a two-cycle engine with the piston nearing TDC. Ignition of the compressed air-fuel mixture occurs approximately at this point. (*Johnson Motors.*)

FIGURE 3.17 As the piston approaches BDC, it uncovers the intake and exhaust ports. Burned gases stream out through the exhaust port, and a fresh charge of air-fuel mixture enters the cylinder, as shown by the arrows. (*Johnson Motors.*)

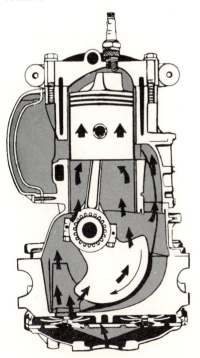
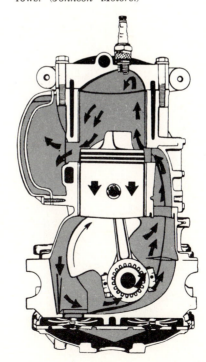

begin to stream out through the exhaust port. At the same time, the intake port, now cleared by the piston, begins to deliver air-fuel mixture, under pressure, to the cylinder. The top of the piston is shaped in such a way as to give the incoming mixture an upward movement. This helps to sweep the burned gases ahead and out through the exhaust port.

After the piston has passed through BDC and starts up again, it passes both ports, thus sealing them off (Fig. 3.18). Now, the fresh air-fuel charge above the piston is compressed and ignited. The same series of events takes place again and continues as long as the engine runs.

3.10 Crankcase Pressure.   The air-fuel mixture is delivered to the cylinder under pressure in most engines; this pressure is applied to the air-fuel mixture in the *crankcase*. (The crankcase is the lower part of the engine which contains the crankshaft.) The crankcase is sealed except for a reed (or leaf) valve at the bottom. The *reed valve* is a flexible, flat metal plate (Fig. 3.19) that rests snugly against the floor

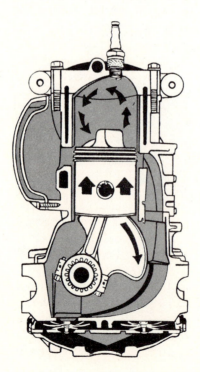

FIGURE 3.18   After the piston passes BDC and moves up again, it covers the intake and exhaust ports. Further upward movement of the piston traps and compresses the air-fuel mixture. (*Johnson Motors.*)

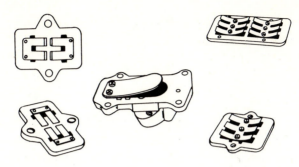

FIGURE 3.19   Reed valves. The blades are flexible so they can move away from the base to which they are attached, or flatten down on the base so as to provide a seal. (*Tecumseh Products Company.*)

of the crankcase. There are holes under the reed valve that connect it with the engine carburetor. When the piston is moving up, a partial vacuum is produced in the sealed crankcase. Atmospheric pressure lifts the reed valve off the holes and pushes air-fuel mixture into the crankcase (Fig. 3.10). After the piston passes TDC and starts down again, pressure begins to build up in the crankcase. This pressure closes the reed valve so that further downward movement of the piston compresses the trapped air-fuel mixture in the crankcase. The pressure which is built up on the air-fuel mixture then causes it to flow up through the intake port into the engine cylinder when the piston moves down enough to clear the intake port (Fig. 3.17).

A disassembled view of a two-cycle engine is shown in Fig. 3.20. Study this illustration carefully; identify the reed valve assembly, cylinder, piston, connecting rod, and other parts. This is an air-cooled engine. Note that the cylinder and head have metal fins to help radiate heat and thus prevent overheating of the engine.

This engine, installed in a lawn mower, is shown in cutaway view in Fig. 3.21. Note that the cylinder is placed horizontally (to the left in the illustration) and that the carburetor is on the opposite side of the crankshaft (to the right in the illustration). A later chapter describes the fuel system.

This engine has a built-in ignition system using a magneto. The magneto is described in a later chapter. It also has a governor which controls engine speed so that the proper engine speed is maintained during engine operation. The governor is described on a later page.

3.11 Transfer Port.   Instead of using a reed valve in the crankcase, some engines have a third, or transfer, port in the cylinder, as shown in Fig. 3.22. In this type of engine, the intake port is cleared by the piston as it approaches TDC. When this happens, the air-fuel mixture pours into the crankcase, filling the partial vacuum left by the upward

FIGURE 3.20 Disassembled view of a one-cylinder, air-cooled engine used on a power lawn mower. (*Lawn Boy Division, Outboard Marine Corporation.*)

1. choke-knob assembly.
2. gas line
3. gas-tank assembly
4. shutoff valve and screen assembly
5. starter pulley
6. flywheel screen
7. flywheel assembly
8. governor assembly
9. magneto assembly
10. air-filter assembly
11. crankshaft
12. carburetor assembly
13. carburetor gasket
14. reed-valve assembly
15. spark plug
16. exhaust sleeve
17. cylinder and sleeve assembly
18. gasket
19. connecting-rod assembly
20. connecting-rod pin
21. piston rings
22. piston
23. crankcase
24. starter rope

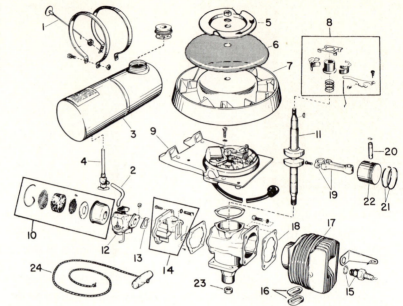

movement of the piston. Then, as the piston moves down, the intake port is cut off by the piston. The air-fuel mixture in the crankcase is compressed, and the other actions then take place as already described.

3.12 Rotary Valves.   In a unit of this type, a rotary valve is built into the crankshaft. This valve lines up with a port in the crankcase

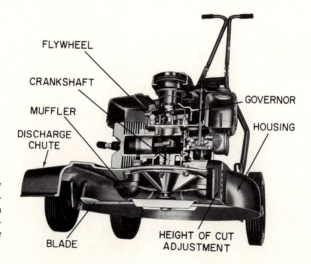

FIGURE 3.21 Cutaway view of a two-cycle engine used in a power lawn mower. (*Lawn Boy Division, Outboard Marine Corporation.*)

FLYWHEEL

CRANKSHAFT

MUFFLER

DISCHARGE CHUTE

GOVERNOR

HOUSING

BLADE

HEIGHT OF CUT ADJUSTMENT

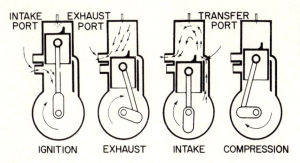

FIGURE 3.22 Actions in a three-port, two-cycle engine. The third port is called the transfer port.

as the piston approaches TDC (Fig. 3.23). Then, as the piston moves toward BDC the port is closed by the rotating crankshaft (Fig. 3.24). Compression takes place in the crankcase as in other two-cycle engines. Then, when the intake port in the cylinder well is uncovered by the piston, the compressed air-fuel mixture can flow from the crankcase into the cylinder.

3.13 Flywheel. The power impulses resulting from the power strokes occur only during one-half a revolution of the crankshaft in two-cycle engines. As the piston passes TDC, the high pressure from the combustion of the air-fuel mixture pushes down on the piston. This push does not last long, however, because as the piston passes the exhaust port, the pressure is relieved. During the rest of the piston, connecting rod, and crankshaft motion, as the piston passes BDC and starts back up, there is no power being produced. It is only the momentum of the moving parts that carries the piston up to TDC so another power stroke can take place. Thus, a one-cylinder engine

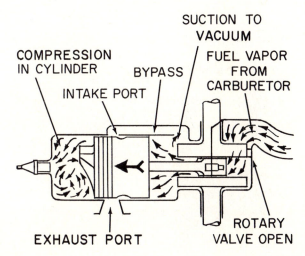

FIGURE 3.23 As the crankshaft rotates, the valve port in the crankshaft lines up with the inlet port in the crankcase to admit air-fuel mixture to the crankcase.

FIGURE 3.24 Further rotation of the crankshaft moves the valve port in the crankshaft away from the inlet port so the valve is closed.

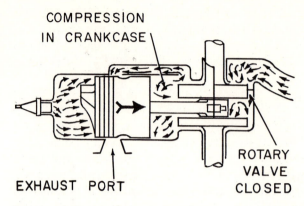

COMPRESSION IN CRANKCASE

EXHAUST PORT

ROTARY VALVE CLOSED

has a tendency to speed up during the power stroke and slow down the rest of the time. To smooth out this speed-up and slow-down action, the engine is equipped with a flywheel.

The flywheel, as shown in Fig. 3.25, is mounted on the end of the crankshaft. It makes use of the property of inertia that all material things have. An object that is moving tries to keep moving. That is inertia. An object that is stationary tries to stay put. That also is inertia. So the flywheel, once it is set in motion, tries to keep moving. Thus, when the engine tries to slow down during the non-power part of the cycle, the flywheel helps to keep it moving. Also, when the power stroke occurs and the engine tries to speed up, the flywheel helps to keep it from suddenly speeding up. You might say that the flywheel gives up energy to keep the engine moving during the non-power time, and then takes in energy when the engine tries to speed up.

In the four-cycle engine, the power stroke occurs only once every four piston strokes, or only one-half of every two revolutions of the crankshaft. This engine must also have a flywheel. In fact, it would seem that the four-cycle engine needs a flywheel even more than a two-cycle engine.

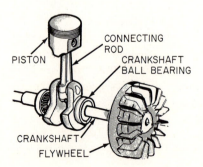

PISTON

CONNECTING ROD

CRANKSHAFT BALL BEARING

CRANKSHAFT

FLYWHEEL

FIGURE 3.25 The flywheel is mounted on the end of the crankshaft and rotates with it.

Multiple-cylinder engines, such as found in automobiles, also use flywheels. The power strokes in these engines are arranged to follow one another in sequence, or overlap, to make a smoother-running engine. But they also need flywheels to further smooth out the engine.

3.14 Engine Cooling. The combustion process in the engine cylinder produces a great deal of heat. Part of this heat is useful — that is, it causes the push on the piston which makes the piston move and the crankshaft rotate. Part of the heat is lost in the hot exhaust gases. And part of the heat passes into the cylinder wall, cylinder head, and piston. This heat must be disposed of to prevent the cylinder walls, head, and piston from getting too hot. Excessive temperatures will cause the oil film on the cylinder wall to fail. That is, without adequate means of disposing of the excess heat, the oil would char, or burn, with the result that the lubricating properties would be lost. The result would be complete engine failure. Without lubrication, the piston rings, piston, and cylinder wall would wear very rapidly. So much added heat would be produced by the dry friction that welds might form. That is, as the piston passed through TDC, the heat might be great enough to cause momentary welding of the rings, or piston, to the cylinder wall. The welds would form at the moment that the piston was stationary and would be caused by actual spot melting of the metal. These welds would then tear loose and cause gouges in the cylinder walls. After a few moments of this sort of action, the piston would probably freeze to the cylinder and there would be complete failure of the engine.

But even before this disaster occurred, there could be serious difficulty in the engine. As the heat accumulated and the cylinder head and spark plug temperature went up, there would soon be a point at which preignition would occur. That is, the spark plug would be so hot that it would ignite the air-fuel mixture prematurely. This condition could also cause engine failure because preignition can knock holes in pistons and cause other internal damage.

To prevent such troubles, there must be some means of getting rid of the excess heat of the cylinder walls, head, and piston. Two different means of doing this are used. One method uses air to carry away the heat. The other uses a liquid, such as water. In the method using air cooling, the cylinder walls and head have fins, as shown in Fig. 3.26. These fins are actually part of the head and cylinder. They greatly increase the outer metal surfaces and thus greatly increase the area from which heat can radiate and thus escape into the surrounding air. To assist in this heat escape, many engines have shrouds and an air fan which forces air over and around the fins. The shrouds are simply metal sheets shaped to fit around the cylinder in such a way

FIGURE 3.26 One-cylinder, air-cooled engine, showing fins in the cylinder and cylinder head. Note the shrouds that direct air over the fins. (*Wisconsin Motor Corporation.*)

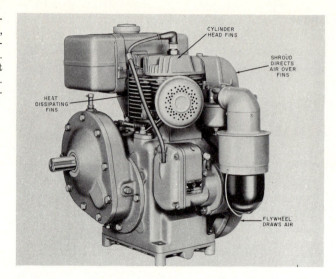

as to force the air to flow close to the fins. The air fan is built into the engine flywheel. The engine is thus cooled by the flow of air around the fins and is therefore called an air-cooled engine. Practically all one-cylinder engines are air cooled. Some multicylinder engines are also air cooled. For example the four-cylinder Volkswagen engine shown in Fig. 3.27 is air cooled.

3.15 Liquid Cooling Systems. The second method of cooling the engine uses a liquid, such as water, mixed with an antifreeze. In the liquid-cooled engine, the liquid is circulated in jackets, or pockets, that surround the cylinder walls and cylinder head. Most multicylinder engines, particularly those used in automobiles, are of the liquid-cooled type. In these, water circulates in spaces around the cylinder walls and through passages in the cylinder head. Antifreeze compounds are added to the water during cold weather to prevent freezing. As the water circulates, it picks up heat from the engine and carries this heat to a radiator. The radiator then cools the water by discharging the heat into the air which is passing through the radiator.

Figure 3.28 shows a cooling system on a V-8 engine. A water pump, driven by the engine fan belt, keeps the water in continuous circulation while the engine is running. Figure 3.28 shows where the water pump is mounted on the cylinder block. The water enters the engine from the bottom of the radiator, passing up through the spaces in the block and head. These spaces are called the *water jackets.*

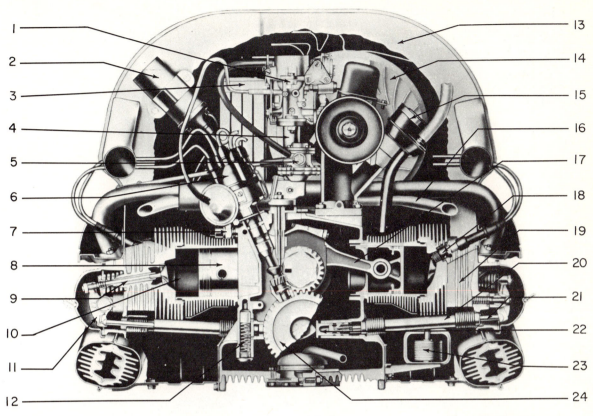

FIGURE 3.27 A cutaway view of a four-cylinder air-cooled automotive engine. (*Volkswagen.*)

1. carburetor
2. ignition coil
3. oil cooler
4. intake manifold
5. fuel pump
6. ignition distributor
7. oil-pressure switch
8. piston
9. valve
10. cylinder
11. rocker arm
12. oil-pressure relief valve
13. fan housing
14. fan
15. oil filler and breather
16. preheating pipe
17. connecting rod
18. spark plug
19. cylinder head
20. valve push-rod tubes
21. push rod
22. cam follower
23. thermostat
24. camshaft

Figure 3.29 shows a disassembled water pump. There is an impeller on a shaft. The impeller has a series of blades. As it revolves, it *impels* (throws) the water outward by centrifugal force; this causes the water to flow through the pump.

The pump is driven by the fan belt, a V-shaped belt, which also drives the generator. The belt is driven by a pulley on the front end of the crankshaft. The engine fan is mounted on the water-pump pulley so that both the fan and the water pump turn together.

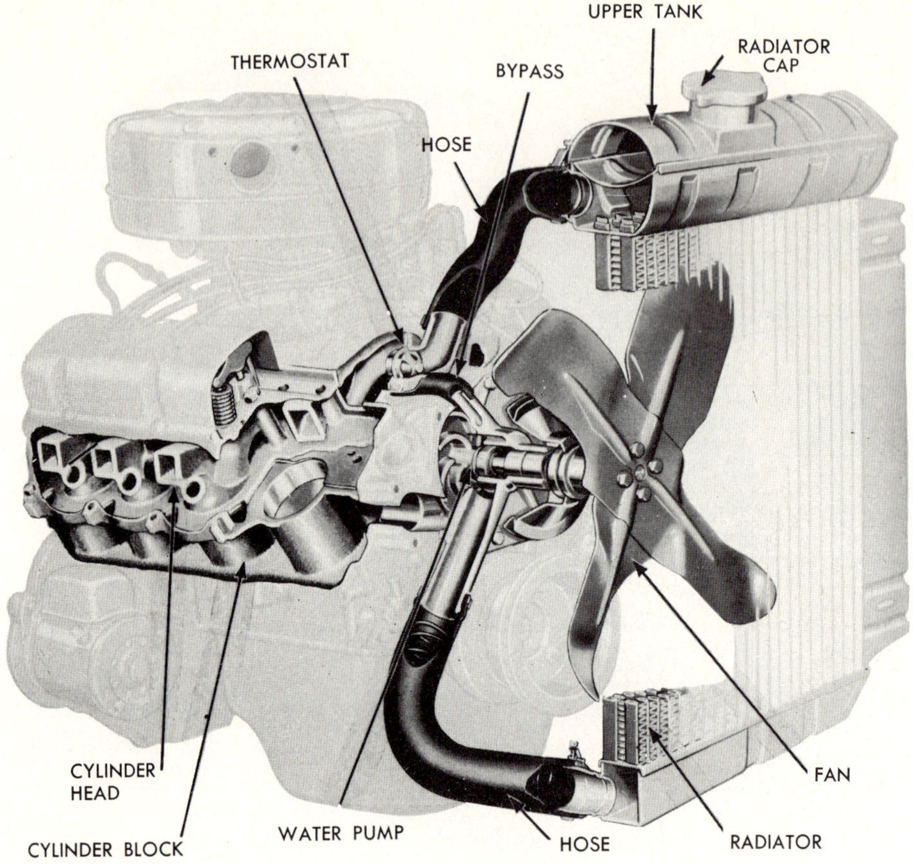

FIGURE 3.28 Cutaway view of a V-8 engine, showing the cooling system. (*Lincoln-Mercury Division, Ford Motor Company.*)

3.16 Radiator.   You can see what a radiator looks like in Fig. 3.30. Figure 3.31 shows one type of radiator in cutaway view. It has a series of water tubes stretching between the upper and lower water tanks. The tubes are surrounded by fins to help radiate the heat from the water in the tubes. Water passes through the tubes and air passes around the tubes between the fins.

3.17 Thermostat.   There is a thermostat located in the passage between the cylinder head and the upper radiator tank (see Fig. 3.28).

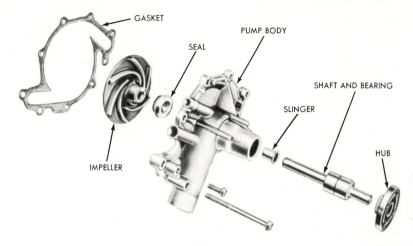

GASKET

SEAL

PUMP BODY

SHAFT AND BEARING

SLINGER

HUB

IMPELLER

FIGURE 3.29 Disassembled view of a water pump. (*Lincoln-Mercury Division, Ford Motor Company.*)

Figure 3.32 shows a thermostat in the open position. In this position, it allows water to pass freely into the radiator. But when the thermostat is closed, it closes off this passage. Now, the water in the engine water jackets cannot leave and the water and engine both heat up very rapidly. This is desirable because the engine should heat up as rapidly as possible after starting. When an engine is operated cold, the lubricating oil flows slowly so that moving parts may not be fully lubricated. This promotes excessive wear. Also, water condensation, as already described, takes place and water drips down into the crankcase. All this is minimized by the thermostat. When the engine is cold,

FIGURE 3.30 Radiator assembly.

FIGURE 3.31 Construction of tube-and-fin radiator core.

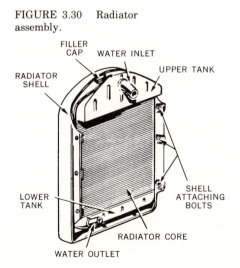

FILLER CAP

WATER INLET

UPPER TANK

RADIATOR SHELL

LOWER TANK

SHELL ATTACHING BOLTS

RADIATOR CORE

WATER OUTLET

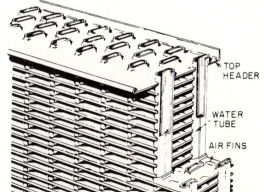

TOP HEADER

WATER TUBE

AIR FINS

a bellows in the thermostat (Fig. 3.32) is shortened up to close the valve so the water cannot circulate to the radiator.

As the engine warms up, a chemical in the bellows evaporates and causes the bellows to expand, thus raising the valve to permit water to circulate between the engine and the radiator. Now, normal water circulation and cooling of the engine can take place.

Thermostats also use wax pellets which expand with heat to open their valves. The sleeve and butterfly thermostats both use wax pellets.

3.18 Radiator Pressure Cap.   Most automotive engines today have pressurized cooling systems. Water boils at about 212° F at sea level. If the pressure is increased, the water will not boil until a higher temperature is reached. Without pressurizing, the cooling system must be designed to prevent the water from reaching 212°. But if the system is pressurized, the water temperature can safely go up to almost 250° without boiling. This higher temperature allows the cooling system to operate more efficiently. Since the water enters the radiator at a higher temperature, the temperature difference between the water and the surrounding air is greater. This causes a greater heat transfer.

The pressure cap contains two spring-loaded valves. One, called the *blowoff valve,* opens if the pressure gets too high to allow the excessive pressure to escape. The other, the vacuum valve, operates

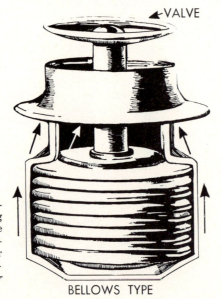

VALVE

FIGURE 3.32   Thermostat for engine cooling system. This bellows type is shown open with arrows indicating water flow past the valve. (*Plymouth Division, Chrysler Motors Corporation.*)

BELLOWS TYPE

when the engine cools off. When this happens, a partial vacuum could form in the cooling system due to condensation of steam. The vacuum valve prevents this by opening to admit air from the outside. If a high vacuum formed, it might cause the radiator to partly collapse, pushed in by atmospheric pressure.

3.19 Temperature Indicator. The temperature indicator tells the driver how hot the engine is getting. An abnormal heat rise is a warning to the driver that something is wrong and he can stop the engine before serious damage results.

CHECKUP

The two-cycle-engine principles you have been studying in the chapter you just completed will help you in your service work on these engines. Once you understand how and why engines work, it becomes easier to put your finger on trouble causes when something goes wrong. The tests that follow will give you a chance to check up on yourself and find out how well you have retained the information in the chapter you have just completed. The tests will also reinforce your memory because you are asked to write down your answers in your notebook. Writing the answers fixes the facts more firmly in your mind.

NOTE: Write your answers in your notebook. Then later you will find that your notebook has become a valuable source of information you can refer to whenever you want an answer to a question.

Completing the Sentences: Test 3. The sentences below are not complete. After each sentence there are several words or phrases, only one of which will correctly complete the sentence. Write each sentence in your notebook, selecting the proper word or phrase to complete it correctly.

1. The engine parts that have the job of providing a good sealing fit between piston and cylinder wall are called (*a*) piston pins; (*b*) piston rings; (*c*) connecting rods.
2. As a rule, the piston in a two-cycle engine has (*a*) one piston ring; (*b*) two piston rings; (*c*) three piston rings.
3. The devices in the engine that change reciprocating motion to rotary motion are the (*a*) piston and rod; (*b*) piston pin, rod, and cap; (*c*) rod and crank.
4. The rod big end is attached to the (*a*) piston pin; (*b*) crankpin; (*c*) piston rings.

5. The sleeve bearing in the connecting-rod big end is made up of (*a*) a solid sleeve; (*b*) a hollow sleeve; (*c*) two bearing halves.
6. The connecting rod is attached to the piston by the (*a*) piston pin; (*b*) rod cap; (*c*) main bearings.
7. The movement of the piston from TDC to BDC is called a (*a*) piston stroke; (*b*) a piston cycle; (*c*) bore and stroke.
8. The valve that holds pressure in the crankcase is called (*a*) an intake valve; (*b*) a pressure valve; (*c*) a reed valve.
9. The devices on the crankshaft that help to balance the weights of the crankpin and connecting rod are called (*a*) balance weights; (*b*) counterweights; (*c*) reaction weights.
10. The purpose of the fins on the cylinder head is to (*a*) strengthen the cylinder; (*b*) hold the heat; (*c*) dispose of the heat.

Written Checkup

In the following, you are asked to write down, in your notebook, the answers to the questions asked or to define certain terms. Writing the answers down will help you to remember them.

1. What is the basic difference between an external combustion engine and an internal combustion engine?
2. What is the purpose of the piston rings and how do they work?
3. Why do four-cycle engines need an extra ring on the piston?
4. What is the purpose of the crank on the crankshaft and how does it work?
5. What is the purpose of the bearings in the engine?
6. Name four places in a two-cycle engine where bearings are used.
7. What are the two basic types of bearings?
8. Describe the actions in a two-cycle engine using a reed valve.
9. Describe the actions in a two-cycle engine using a transfer port.
10. What is the purpose of the flywheel?
11. What are the two basic types of cooling system?
12. Explain how a liquid cooling system for an engine works.

Four-Cycle Engine Operation

4

We will now look at the construction and operation of four-cycle engines. Most automotive engines are of the four-cycle type, and many small engines also operate on the four-cycle principle. The basic difference between the two-cycle and four-cycle engines is in the manner in which the air-fuel mixture is introduced into the cylinder, and the way in which the burned gases are removed from the cylinder after the power stroke of the piston.

4.1 Comparing the Two-Cycle and Four-Cycle Engines. The four-cycle engine is very similar in many ways to the two-cycle engine. In both engines, a piston moves up and down in the cylinder. The piston is attached to a crank on the crankshaft by a connecting rod. When ignition of the compressed air-fuel mixture takes place, the high pressure drives the piston down. This push, carried through the crankshaft causes the crankshaft to rotate.

To this point, the actions are similar in both engines. However, in the two-cycle engine, the air-fuel mixture is admitted to the cylinder, and the burned gases exit from the cylinder, through openings, or valve ports, in the cylinder wall. See Figs. 3.15 to 3.18. Also, in the two-cycle engine, air-fuel mixture is compressed in the cylinder every time the piston moves up. Every time the piston nears TDC, there is combustion and high pressure which pushes the piston down. Only two piston strokes are required to complete the cycle of engine operation in the two-cycle engine. Actually, the proper name is two-stroke-cycle engine because there are two piston strokes in the cycle. We have mentioned this previously. The proper name of the four-cycle engine is four-stroke-cycle engine because it takes four piston strokes to complete the cycle.

The four-cycle engine does not have valve ports in the cylinder wall. Instead, this engine has two openings at the top of the cylinder. These openings are plugged, part of the time, by movable metal plugs called valves. Figure 4.1 shows typical valves for a four-cycle engine. These valves are operated by means of a gear-and-cam arrangement.

53

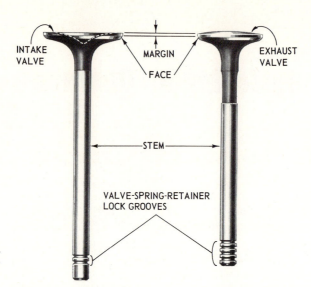

INTAKE
VALVE

MARGIN

FACE

EXHAUST
VALVE

STEM

VALVE-SPRING-RETAINER
LOCK GROOVES

FIGURE 4.1 Typical engine valves. (*Chrysler Motors Corporation.*)

They move up and down in valve guides in the cylinder block. One of the valves operates to allow air-fuel mixture to enter the cylinder. The other valve operates to allow the burned gases to escape from the cylinder.

4.2 Operating the Valves.   Figure 4.2 shows a typical arrangement for operating a valve in a small engine. The valve moves up and down in a valve guide which is part of the cylinder block. There is a valve spring that puts tension on the valve and tries to keep the valve closed, that is, seated on the valve seat, or port, in the cylinder block. The valve spring is held between the cylinder block and a spring retainer. The spring retainer is attached to the valve stem by a retainer lock which fits into grooves in the valve stem. Below the valve stem is a valve lifter or valve tappet as it is also called. The valve lifter moves up and down in a bore, or hole, in the cylinder block. The valve lifter rests on a cam which has one high spot, or lobe. The cam rotates as the crankshaft rotates. The two are geared together. Figure 4.3 shows a gearing arrangement for a small four-cycle engine. As the camshaft rotates, the cam lobe moves around under the valve lifter, causing it to be pushed upward, as shown in Fig. 4.4. This upward push overcomes the valve-spring tension so that the valve is raised off the valve seat. The valve is then open, and gas can pass through the opening between the valve seat and valve. If the valve is the intake valve, it is air-fuel mixture from the carburetor that passes through the valve opening on its way to the cylinder. If it is the exhaust valve that opens,

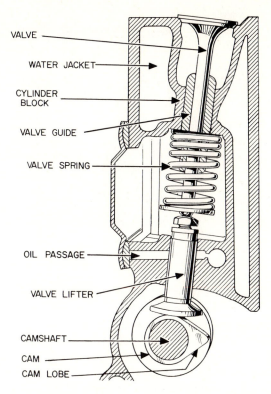

VALVE

WATER JACKET

CYLINDER BLOCK

VALVE GUIDE

VALVE SPRING

OIL PASSAGE

VALVE LIFTER

CAMSHAFT

CAM

CAM LOBE

FIGURE 4.2 Valve mechanism used in an L-head engine. The valve is raised up off its seat with every camshaft rotation.

then it is the exhaust gases—that is, the gases left after the air-fuel mixture has burned—which pass through the valve opening on their way out from the engine cylinder.

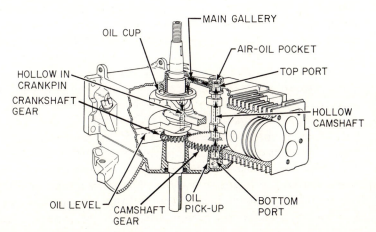

MAIN GALLERY

OIL CUP

AIR-OIL POCKET

TOP PORT

HOLLOW IN CRANKPIN

CRANKSHAFT GEAR

HOLLOW CAMSHAFT

OIL LEVEL

CAMSHAFT GEAR

OIL PICK-UP

BOTTOM PORT

FIGURE 4.3 Gearing arrangement in a small four-cycle horizontal engine for driving the camshaft from the crankshaft. This illustration also shows the flow of lubricating oil to the camshaft, crankshaft, and connecting-rod bearings. (*Kohler Company.*)

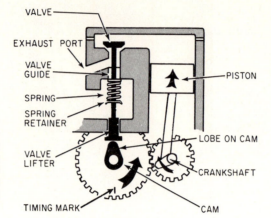

FIGURE 4.4 As the cam-shaft is driven by the crankshaft, the cam lobe moves up under the valve tappet, and this forces the valve to move up off its seat.

Then, as the piston continues its movement and the crankshaft continues to rotate, the gears on the crankshaft rotate the camshaft so that the lobe on the cam moves out of the way of the valve lifter. Now, the spring on the valve forces the valve to close so the opening is sealed off, as shown in Fig. 4.5.

You will notice that the gear on the camshaft is twice as large as the gear on the crankshaft. There is a reason for this; the camshaft should rotate just half as fast as the crankshaft. This makes the cam-shaft rotate once for every two times the crankshaft turns. We will explain why this is necessary later.

4.3   Engine Operation.   We are now ready to take a closer look at the operation of the four-stroke-cycle engine. Figures 4.6 through 4.9 illustrate these four strokes. To start with, the intake valve has been pushed up off its seat by a cam lobe on the rotating camshaft as the piston is moving down. A mixture of air and fuel is being taken into the cylinder, flowing past the intake valve. This is the intake stroke, shown in Fig. 4.6.

Then, as the piston passes BDC, the intake valve closes. Now, the piston starts up on the compression stroke, compressing the air-fuel mixture into the top of the cylinder. See Fig. 4.7.

Next, as the piston nears TDC on the compression stroke, an electric spark takes place at the spark plug. This sets fire to, or ignites, the compressed air-fuel mixture. Combustion and a pressure rise result, forcing the piston downward on the power stroke. The downward push on the piston may total as much as a ton in a modern small engine. This powerful push is carried through the connecting rod to a crank on the engine crankshaft (Fig. 4.8). The electric ignition

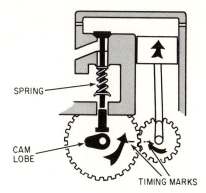

FIGURE 4.5 Further rotation of the crankshaft and camshaft moves the lobe out from under the valve tappet, and this allows the spring to pull the valve down and reseat it.

system, which produces the spark at the spark plug, will be explained later.

Finally, the fourth stroke in the four-stroke-cycle occurs. This is the exhaust stroke. As the piston nears BDC on the power stroke, the exhaust valve opens. Now, as the piston moves up on the exhaust stroke, the burned gases in the cylinder are forced out, as shown in Fig. 4.9.

As the piston nears TDC on the exhaust stroke, the intake valve opens. Then, after TDC, the exhaust valve closes, and the whole cycle of events is repeated once again. The cycle is repeated continuously as long as the engine runs.

A completely disassembled view of the engine, shown in Figs. 4.6 to 4.9, is illustrated in Fig. 4.10. You will be able to identify all the parts we have mentioned—the piston, rings, connecting rod, crankshaft, and valves—if you will study this picture for a moment.

You can now see why the gear on the camshaft has to be twice as large as the gear on the crankshaft, as shown in Figs. 4.3 to 4.5. Each valve must open once while the crankshaft is turning two times. That is, each valve is open for only one piston stroke. Since there are four piston strokes in a complete cycle, this means that a valve is open only one-fourth of the total running time.

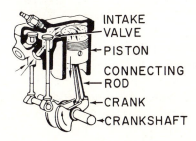

FIGURE 4.6 The intake stroke of a four-cycle engine. The intake valve has closed and the piston is moving down, drawing air-fuel mixture into the cylinder as shown. (*Clinton Engines Corporation.*)

FIGURE 4.7 Compression stroke. Both valves are closed, and the piston is moving upward, compressing the mixture. (*Clinton Engines Corporation.*)

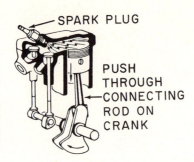

FIGURE 4.8   Power stroke. The ignition system produces a spark at the spark plug that ignites the mixture. As it burns the high pressure created pushes the piston down. (*Clinton Engines Corporation.*)

4.4 Piston Rings.   We mentioned in the previous chapter, in Sec. 3.2, that two-cycle engines use two piston rings on the piston while the four-cycle engine uses three or more rings. We explained how the upper rings—the compression rings—work to hold the pressures in the combustion chamber. We also noted how oil is mixed with gasoline for the two-cycle engine and that this oil provides lubrication of the piston rings and piston.

In the small four-cycle engine, a different method of lubricating the cylinder wall, piston, and rings, is used. A supply of oil is kept in the bottom of the crankcase and this oil is splashed or pumped around so that droplets hit the cylinder wall and keep it oiled. At the same time, some of the droplets hit the valves and valve tappets, permitting them to move up and down easily on films of oil. The oil also covers the bearings in the engine so they are adequately lubricated. Figure 4.3 shows the lubrication system on one small four-cycle, horizontal engine.

There is quite a bit of oil splashed on the cylinder wall. So much that the two compression rings would pass a good deal of it. This oil could then work up into the combustion chamber where it would be burned. The burned oil would leave a carbon residue that would soon clog the valves and spark plug, preventing them from working properly. The engine would therefore begin to lose power and would soon stop working altogether. To prevent this, the piston on four-cycle engines is equipped with a third ring, called the oil-control ring, as shown in Fig. 3.5. Its purpose is to scrape excess oil off the cylinder walls on every downstroke of the piston. It scrapes the oil off so the oil drops back down into the crankcase instead of working its way up into the combustion chamber.

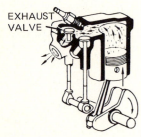

FIGURE 4.9   Exhaust stroke. The exhaust valve has opened and the piston is moving upward, forcing the burned gases from the cylinder, as shown by the arrows. (*Clinton Engines Corporation.*)

4.5 Engine Bearings.   We have already described the various kinds of bearings used in small engines (see Sec. 3.5). You will find both the sliding and rolling types of bearings in small four-cycle engines. We

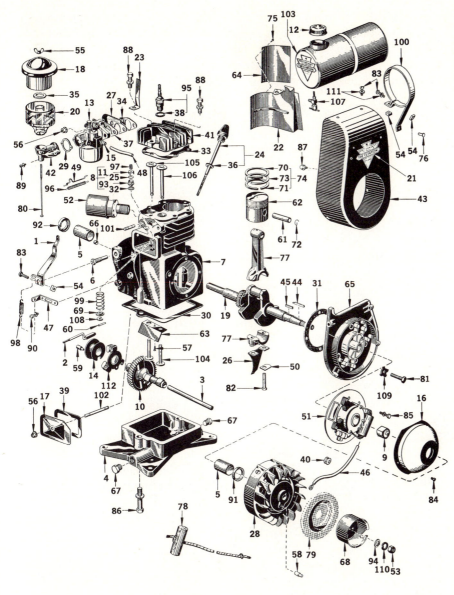

FIGURE 4.10 Exploded view of a typical single-cylinder, four-cycle, air-cooled engine. Only the main parts are identified. (*Clinton Engines Corporation.*)

1. Arm-governor throttle—used on type *B*
2. Arm and weight assembly governor
4. Base engine
5. Bearing main (1-bearing plate, 1-block)
6. Bearing shaft, governor throttle
7. Block assembly cylinder
8. Breather assembly
9. Cam-breaker points
10. Camshaft
14. Collar assembly governor
16. Cover-dust breaker points
22. Deflector air
25. Disk breather
26. Distributor oil
61. Pin wrist
62. Piston standard
63. Plate baffle, crankcase
64. Plate name, Clinton Engines
65. Plate bearing
69. Retainer-valve spring
74. Ring set-standard (1/16 oil ring)
77. Rod assembly-connecting
91. Seal oil (bearing plate)
92. Seal oil (cylinder block)
93. Seat breather
96. Spring backlash, governor
97. Spring-breather hold down
99. Spring valve
104. Tappet valve

will describe the method of lubricating bearings, as well as other moving parts in the engine, in a following chapter on oil and lubricating systems.

4.6 Flywheel. Four-cycle engines require flywheels, and we have described flywheels and their purpose in a previous section (Sec. 3.13).

Flywheels not only serve to keep the engine running more smoothly, as previously explained, but they may also carry the magnets that are part of the magneto ignition system, and also part of the alternator that produces current to charge the battery. We will have more to say about these devices in following chapters.

4.7 Valve Timing.   We mentioned that the valves open and close, not at TDC, or at BDC, but some time before or after the piston reaches the upper or lower limit of travel. There is a reason for this. Look at the intake valve, for example. It normally opens several degrees of crankshaft rotation before TDC on the exhaust stroke. That is, the intake valve begins to open before the exhaust stroke is finished. This gives the valve enough time to reach the fully open position before the intake stroke begins. Then, when the intake stroke starts, the intake valve is already wide open and air-fuel mixture can start to enter the cylinder immediately. Likewise, the intake valve remains open for quite a few degrees of crankshaft rotation after the piston has passed BDC at the end of the intake stroke. This allows additional time for the air-fuel mixture to continue to flow into the cylinder. The fact that the piston has already passed BDC and is moving up on the compression stroke while the intake valve is still open does not affect the movement of air-fuel mixture into the cylinder. Actually, air-fuel mixture is still flowing in as the intake valve starts to close.

The reason for this is that the air-fuel mixture has inertia. That is, it attempts to keep on flowing after it once starts through the carburetor and into the engine cylinder. The momentum of the mixture then keeps it flowing into the cylinder even though the piston has started up on the compression stroke. This packs more air-fuel mixture into the cylinder and results in a stronger power stroke. In other words, this improves volumetric efficiency (see Sec. 5.10).

For a somewhat similar reason, the exhaust valve opens well before the piston reaches BDC on the power stroke. As the piston nears BDC, most of the push on the piston has ended and nothing is lost by opening the exhaust valve toward the end of the power stroke. This gives the exhaust gases additional time to start leaving the cylinder so that exhaust is well started by the time the piston passes BDC and starts up on the exhaust stroke. The exhaust valve then stays open for some degrees of crankshaft rotation after the piston has passed TDC and the intake stroke has started. This makes good use of the momentum of the exhaust gases. They are moving rapidly toward the exhaust port, and leaving the exhaust valve open for a few degrees after the intake stroke starts gives the exhaust gases

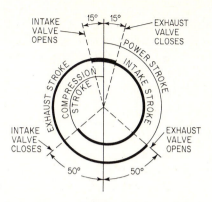

FIGURE 4.11 Intake- and exhaust-valve timing in a typical small engine. The complete cycle of events is shown as a 720° spiral, which represents two complete crankshaft revolutions. The timing of valves differs for different engines.

some additional time to leave the cylinder. This allows more air-fuel mixture to enter on the intake stroke so that a stronger power stroke results. That is, it improves volumetric efficiency (see Sec. 5.10).

Actual timing of the valves varies with different four-cycle engines, but a typical example for an engine is shown in Fig. 4.11. Note that the intake valve opens 15° of crankshaft rotation before TDC on the exhaust stroke, and stays open until 50° of crankshaft rotation after BDC on the compression stroke. The exhaust valve opens 50° before BDC on the power stroke and stays open 15° after TDC on the intake stroke. This gives the two valves an overlap of 30° at the end of the exhaust stroke and beginning of the compression stroke.

4.8 Automatic Compression Release. Cranking an engine is not the easiest thing in the world. To pull the engine through the compression stroke requires some effort, either muscle power or starting-motor power. One way to reduce this effort is to partly release compression during cranking. One method of doing this is shown in Fig. 4.12. The mechanism involved consists of a pair of flyweights on the camshaft drive gear. When the engine is not running, the flyweights are held in their inner position by springs, as shown to the left in Fig. 4.12. In this position, a tang on the end of one of the flyweights has moved out of a notch in the base circle of the exhaust cam, as shown to the lower left. Now, when the engine is cranked for starting, this tang prevents the exhaust valve from closing completely. Every time the base circle of the cam comes around under the valve tappet of the exhaust valve, it prevents the tappet from moving all the way down. With the exhaust valve held partly open, some of the compression pressure is relieved.

After the engine starts and engine speed increases, centrifugal force acting on the two flyweights forces them to move out into the running position, as shown to the right in Fig. 4.12. This movement

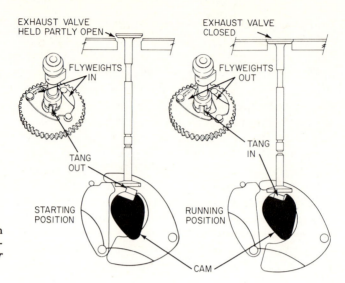

EXHAUST VALVE
HELD PARTLY OPEN

FLYWEIGHTS
IN

TANG
OUT

STARTING
POSITION

EXHAUST VALVE
CLOSED

FLYWEIGHTS
OUT

TANG
IN

RUNNING
POSITION

CAM

FIGURE 4.12 Operation of the automatic compression release. (*Kohler Company*.)

allows the tang on the end of one of the flyweights to move into the notch in the base circle of the cam, as shown to the lower right in Fig. 4.12. Now, when the base circle of the cam for the exhaust valve comes around under the valve tappet, the valve tappet can come all the way down and thus allow the exhaust valve to close completely. Now, engine operation continues in a normal manner as long as the engine speed is maintained. However, if the engine is stopped, then the springs will cause the flyweights to move into the starting position, as shown to the left in Fig. 4.12 in readiness for another starting cycle.

Not only does the automatic compression release reduce the amount of effort required to start the engine, but it also reduces the time required to start, according to the engine manufacturer using this device. For one thing, with reduced compression, less voltage is required to fire the spark plugs. This seems obvious because the higher the compression, the higher the voltage must go in order to force a spark to jump through the denser air-fuel mixture.

4.9 I-Head Engine. The type of engine described above and shown in Figs. 4.2 to 4.10 is called an L-head engine because the cylinder and combustion chamber are in the shape of an inverted L (Fig. 4.13). Most small four-cycle engines are of this type. They have the valves located in the cylinder block. Some small four-cycle engines, and almost all automotive engines, are of the I-head, or overhead-valve type. In this type of engine, the valves are located overhead, in the cylinder head. Figure 4.14 shows a cutaway view of this type of engine. Figure 4.15 shows the essential parts required to operate the valves. Note that the I-head engine requires two more parts than the L-head engine—push rods and rocker arms. The rocker arms are held in place on a shaft, or on ball studs. With either arrangement, the rocker arms are free to rock back and forth, just like a teeter-totter. When the cam lobe moves around under the valve lifter, the lifter is raised and this pushes up on the push rod. The push rod causes the rocker arm to rock so that the rocker arm pushes down on the end of the valve stem. As a result, the valve is pushed down off the valve seat. That is, the valve opens. This type of engine is often called a push-rod engine. As we mentioned, most small engines are of the L-head type because they are simpler in construction and easier to service. You may find, however, some small engines with the valves in the cylinder head.

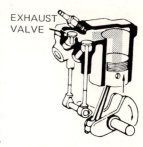

FIGURE 4.13 The L-head engine gets its name from the shape of the cylinder and combustion chamber which form an inverted "L" in the engine shown here. (*Clinton Engines Corporation.*)

4.10 T-Head Engine. A few small engines have been made with the T-head arrangement. See Fig. 4.16, which compares the L-head, T-head, and I-head arrangements. In the T-head arrangement, two camshafts are required, one for each side of the engine. One of the valves is the intake valve, the other the exhaust valve. As you can see, the combustion chamber and cylinder form a "T."

FIGURE 4.14 Partial cutaway view of a four-cylinder, in-line, overhead-valve engine. (*Chevrolet Motor Division of General Motors Corporation.*)

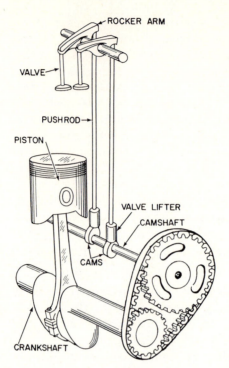

FIGURE 4.15   Valve-operating mechanism for an I-head, or overhead-valve engine.

4.11 Comparison of the Two-Cycle and Four-Cycle Engines.   We have previously explained that it takes two revolutions of the crankshaft to complete the four strokes in a four-stroke-cycle engine. The first half revolution, or 180°, the piston is moving down on the intake stroke. The next half revolution, the piston is moving up on the compression stroke. The third half revolution of the crankshaft, the piston is moving down on the power stroke. The fourth half revolution, the piston is moving up on the exhaust stroke. Figures 4.6 through 4.9 show these four strokes.

Most larger internal-combustion engines and practically all automotive engines are of the four-stroke-cycle type. Every fourth piston stroke in each cylinder is a power stroke.

In the two-stroke-cycle engine (or two-cycle engine as it is commonly called), a power stroke occurs every two piston strokes. That is, every downward movement of the piston is a power stroke. In effect, the intake and compression strokes are combined. Also, the power and exhaust strokes are combined. We have described how the two-cycle engine works in Chap. 3.

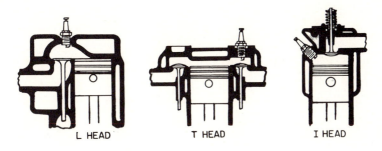

L HEAD   T HEAD   I HEAD

FIGURE 4.16 Sectional views of three types of engine heads—showing three valve arrangements.

You might think that, because the two-cycle engine has twice as many power strokes as a four-cycle engine (Fig. 4.17), it would produce twice as much horsepower as a four-cycle engine of the same size, running at the same speed. However, this is not true. In the two-cycle engine, when the intake and exhaust ports have been cleared by the piston, there is always some mixing of the fresh charge and the burned gases. Not all the burned gases get out, and this prevents a larger fresh charge from entering. Therefore, the power stroke that follows is not as powerful as it could be if all the burned gases were exhausted and a full charge of air-fuel mixture entered. In the four-cycle engine, nearly all the burned gases are forced from the combustion chamber by the upward-moving piston. And a comparatively full charge of air-fuel mixture can enter because a complete piston stroke is devoted to the intake of the mixture (contrasted with only part of a stroke on the two-cycle engine). Therefore, the power stroke in the four-cycle engine produces more power.

However, the two-cycle engine is widely used as a power plant for lawn mowers, motor boats, snow removers, model airplanes, motor scooters, power saws, and other such equipment. These engines are often air cooled. Because they have no valve train or water cooling system, they are relatively simple in construction and light in weight. These are desirable characteristics for engines used on small, lightweight equipment that must be handled and moved around.

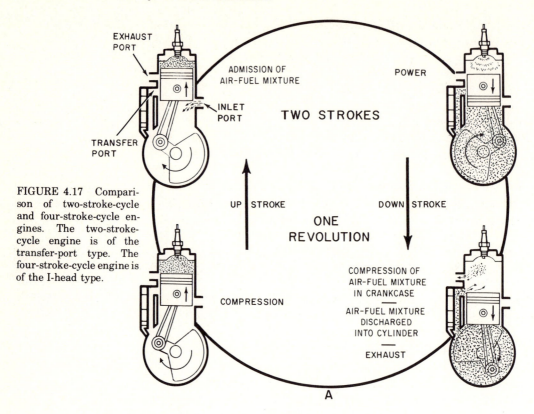

FIGURE 4.17 Comparison of two-stroke-cycle and four-stroke-cycle engines. The two-stroke-cycle engine is of the transfer-port type. The four-stroke-cycle engine is of the I-head type.

4.12 Operating Position of Crankshaft. We have discussed different classifications of small engines: two-cycle or four-cycle; I-head, L-head, or T-head; air-cooled or water-cooled. Small two-cycle engines can be classified in another way, by the operating position of the crankshaft. There are three basic positions.

1. Vertical, as shown in Fig. 3.21
2. Horizontal, as shown in Fig. 4.18
3. Multiple positions as would be required with a chain saw (Fig. 4.19)

4.13 How to Tell a Two-Cycle from a Four-Cycle Engine. The expert can usually tell at a glance whether an air-cooled engine is of the two-cycle or four-cycle type. The four-cycle engine has an oil sump and an oil-filler plug. The two-cycle engine does not. The four-cycle engine requires oil drains and refills periodically, just as automobiles do. In the two-cycle engine, the oil is added to the gasoline so that a mixture of gasoline and oil passes through the carburetor and enters the crankcase with the air. The gasoline evaporates and is burned in the cylinder to produce power. Some of the oil burns too, but some of it remains in the crankcase as mist. It covers the piston, crankshaft, bearings, and cylinder wall to provide adequate lubrication of all parts.

HORIZONTAL
CRANKSHAFT

FIGURE 4.18 Some small engines have a horizontal crankshaft such as this one used in a riding mower.

Another distinguishing feature is that in the four-cycle engine, the muffler is installed at the head end of the cylinder, at the exhaust-valve location. See Fig. 4.20. The muffler on the two-cycle engine is installed toward the middle of the cylinder, at the exhaust-port location. See Fig. 3.21.

Also, a look at the nameplate will tell you whether the engine is a two-cycle or a four-cycle unit. If the nameplate tells you to mix oil with the gasoline, you know it is a two-cycle engine. If the nameplate mentions the type of oil and the crankcase capacity, or similar data, you know it is a four-cycle engine.

CHECKUP

The four-cycle-engine principles you have been studying in the chapter you have just completed explain the basic differences between the two-cycle and the four-cycle engine. You have learned about the valves used in four-cycle engines and how they operate. Knowing these fundamentals will be of great help to you when you begin servicing four-cycle engines because you will know why and how they work. The following questions will not only give you a chance to check up on how well you understand and remember these fundamentals, but also will help you to remember them better. The act of writing down the answers to the questions will fix the facts more firmly in your mind.

FIGURE 4.19   A chain saw must be used in many different positions. (*Homelite Division of Textron, Inc.*)

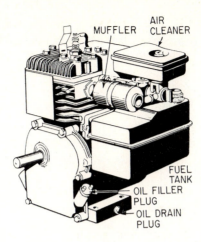

MUFFLER AIR CLEANER

FUEL TANK
OIL FILLER PLUG
OIL DRAIN PLUG

FIGURE 4.20 The muffler is installed at the head end of the cylinder in four-cycle engines. (*Briggs and Stratton Corporation.*)

Completing the Sentences: Test 4. The sentences below are not complete. After each sentence there are several words or phrases, only one of which will correctly complete the sentence. Write each sentence in your notebook, selecting the proper word or phrase to complete it correctly.

1. The four-cycle engine uses (*a*) two valves; (*b*) reed valves; (*c*) transfer ports; (*d*) rotary valves.
2. The camshaft rotates once for every (*a*) two crankshaft revolutions; (*b*) crankshaft revolution; (*c*) one-half crankshaft revolution.
3. The valve is pushed off its seat by the (*a*) crankshaft; (*b*) valve spring; (*c*) camshaft lobe.
4. The two valves used in the four-cycle engine are called the (*a*) exhaust valve and outlet valve; (*b*) intake valve and exhaust valve; (*c*) intake valve and outlet valve.
5. The four strokes of the piston in the four-cycle engine are (*a*) intake, compression, power, exhaust; (*b*) intake, power, outlet, expansion; (*c*) intake, expansion, ignition, power.
6. The purpose of the third ring on the piston in four-cycle engines is to (*a*) hold compression; (*b*) lubricate cylinder wall; (*c*) control oil.
7. The intake valve closes (*a*) before BDC; (*b*) after BDC; (*c*) at BDC.
8. The exhaust valve opens (*a*) before BDC; (*b*) after BDC; (*c*) at BDC.
9. The purpose of the automatic compression release is to (*a*) increase compression ratio; (*b*) make cranking easier; (*c*) reduce top speed.

10. The I-head engine requires two more parts than the L-head four-cycle engine; these are (*a*) lifter and retainer; (*b*) rocker arm and lifter; (*c*) rocker arm and push rod.

Written Checkup

In the following, you are asked to write down, in your notebook, the answers to the questions asked or to define certain terms. Writing the answers down will help you to remember them.

1. Explain the basic differences between the two-cycle and the four-cycle engine.
2. Explain how the valves in the four-cycle engine are operated.
3. Why must the camshaft revolve only half as fast as the crankshaft?
4. What are the four strokes in the four-cycle engine? Explain what happens during each stroke.
5. Why must the four-cycle engine have an extra piston ring?
6. What is meant by valve timing? Describe the timing of the valves in a typical engine.
7. Why does the intake valve stay open after the piston passes BDC?
8. Why does the exhaust valve open before the piston reaches BDC?
9. Explain the purpose of the automatic compression release and how it works.
10. Explain the differences between the L-head and the I-head engine.
11. What are the three basic operating positions of the crankshaft in small engines?
12. Explain the various ways you can tell a two-cycle engine from a four-cycle engine.

Engine Measurements

5

Before we describe engine ignition, fuel, and lubricating systems, let us take a look at the various ways in which engines are measured. These include not only physical measurements, such as cylinder diameter, length of piston stroke, and so on, but also engine performance measurements such as torque and horsepower.

5.1 Work. Work is the moving of an object against an opposing force. The object is moved by a push, a pull, or a lift. For example, when a weight is lifted, it is moved upward against the pull of gravity. Work is done on the weight. Also, when a coil spring is compressed, work is done on the spring (Fig. 5.1).

Work is measured in terms of the distance and force. If a 5-lb weight is lifted off the ground 1 ft, the work done on the weight is 5 ft-lb (foot-pounds), or 1 times 5. If the 5-lb weight is lifted 2 ft, the work done is 10 ft-lb:

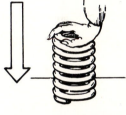

FIGURE 5.1   When a spring is compressed, work is done on that spring, and energy is stored in it.

Distance times force equals work

5.2 Energy. Energy is the ability, or capacity, to do work. When work is done on an object, energy is stored in that object. Lift a 20-lb weight 4 ft and you have stored up energy in the weight. The weight can do 80 ft-lb of work. If a spring is compressed, energy is stored in it, and it can do work (Fig. 5.2).

5.3 Power. Work can be done slowly, or it can be done rapidly. The rate at which work is done is measured in terms of power. A machine that can do a great deal of work in a short time is called a *high-powered* machine:

Power is the rate, or speed, at which work is done

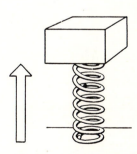

FIGURE 5.2   When the spring is released, it can do work on another body, lifting a weight against the force of gravity, for instance.

5.4 Engine Power. Engines are rated in terms of the amount of power they can produce as compared to the power that a horse can

71

produce. A horse theoretically can produce 33,000 foot-pounds (ft-lb) of work a minute. Figure 5.3 shows what this means. If the horse walks 165 ft in 1 min, it will be lifting the 200-lb weight 165 ft. Work is the moving of any object against an opposing force. The opposing force here is gravity, pulling down on the weight with a force of 200 lb. The amount of work done in 1 min is therefore 200 lb times 165 ft or 33,000 ft-lb which is 1 horsepower (hp). The formula for horsepower is

$$hp = \frac{\text{ft-lb per min}}{33,000} = \frac{L \times W}{33,000 \times t}$$

where hp = horsepower
L = length, in feet, through which W is moved
W = force, in pounds, that is exerted through distance L
t = time, in minutes, required to move W through L

The typical two-cycle engine used in power mowers will produce several horsepower. A 3-hp engine, for example, can theoretically deliver the same amount of power as three horses. It must be remembered, however, that for a 3-hp engine to actually produce 3 hp, it must be in good condition with all parts working properly.

5.5 Torque.   Torque is twisting or turning effort. You apply torque to the top of a screw-top jar when you loosen it (Fig. 5.4). You apply torque to the steering wheel when you take a car around a turn. The engine applies torque to the wheels to make them rotate.

Torque, however, must not be confused with power. Torque is turning effort which may or may not result in motion. Power is some-

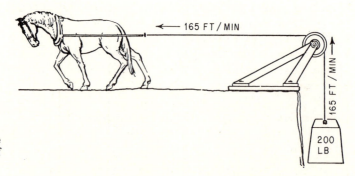

FIGURE 5.3   One horse can do 33,000 ft-lb of work a minute.

thing else again. It is the rate at which work is being done, and this means that something must be moving.

Torque is measured in pound-feet (or lb-ft, not to be confused with ft-lb of work). For example, if you pushed on a windlass crank with a 20-lb push and if the crank were $1\frac{1}{2}$ ft long (from center to handle), you would be applying 30 lb-ft of torque to the crank (Fig. 5.5). You would be applying this torque regardless of whether or not the crank was turning, just so long as you continued to apply the 20-lb push to the crank handle.

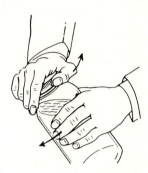

FIGURE 5.4 Torque, or twisting effort, must be applied to loosen and remove the top from a screw-top jar.

5.6 Friction.   Friction is the resistance to motion between two objects in contact with each other. If you put this book on a table and then pushed the book, you would find that it took a certain amount of push. If you put a second book on top of the book, you would find that you had to push harder to move the two books on the table top. Thus, friction, or resistance to motion, increases with the load. The higher the load, the greater the friction. There are three classes of friction: dry, greasy, and viscous.

1. Dry Friction.   This is the resistance to relative motion between two dry objects, for instance, a board being dragged across a floor.

2. Greasy Friction.   This is the friction between two objects thinly coated with oil or grease. In an engine, greasy friction may occur in an engine on first starting. Most of the lubricating oil may have drained away from the bearing surfaces and from the cylinder walls and piston rings. When the engine is started, only the small amount of oil remaining on these surfaces protects them from undue wear. Of course, the lubricating system quickly supplies additional oil, but before this happens, greasy friction exists on the moving surfaces. The lubrication between the surfaces where greasy friction exists is not sufficient to prevent wear. This is the reason automotive engineers say that initial starting and warm-up of the engine is hardest on the engine and wears it the most.

20-LB PUSH

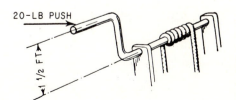

1 1/2 FT

FIGURE 5.5   Torque is measured in pound-feet (lb-ft) and is calculated by multiplying the push by the crank offset, or the distance of the push from the rotating shaft.

FIGURE 5.6 Shaft rotation causes layers of clinging oil to be dragged around with it. The oil moves from the wide clearance *A* and is wedged into the narrow clearance *B*, thereby supporting the shaft weight *W* on an oil film. The clearances are exaggerated in the illustration.

3. Viscous Friction. "Viscosity" is a term that refers to the tendency of liquids, such as oil, to resist flowing. A heavy oil is more viscous than a light oil and flows more slowly (has a higher viscosity, or higher resistance to flowing). Viscous friction is the friction, or resistance to relative motion, between adjacent layers of liquid. In an engine bearing supplied with sufficient oil, layers of oil adhere to the bearing and shaft surfaces. In effect, layers of oil clinging to the shaft are carried around by the rotating shaft. They wedge between the shaft and the bearing (Fig. 5.6). The wedging action lifts the shaft so that the oil itself supports the weight, or load. Now, since the shaft is supported ("floats") on layers of oil, there is no metal-to-metal contact. However, the layers of oil must move over each other, and it does require some energy to make them do so. The resistance to motion between these oil layers is called viscous friction.

5.7 Bore and Stroke. The size of an engine cylinder is referred to in terms of the bore and stroke. The bore is the diameter of the cylinder. The stroke is the distance the piston travels from BDC (bottom dead center) to TDC (top dead center). (See Fig. 5.7.) The bore is always mentioned first. For example, in a 4- by $3\frac{1}{2}$-in. cylinder, the diameter, or bore, is 4 in., and the stroke is $3\frac{1}{2}$ in. These measurements are used to figure the piston displacement.

5.8 Piston Displacement. Piston displacement is the volume that the piston displaces as it moves from BDC to TDC. Piston displace-

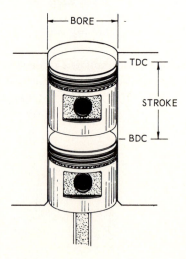

FIGURE 5.7 Bore and stroke of an engine cylinder.

ment of a 4- by $3\frac{1}{2}$-in. cylinder, for example, would be the volume of a cylinder 4 in. in diameter and $3\frac{1}{2}$ in. long, or

$$\frac{\pi \times D^2 \times L}{4} = \frac{3.1416 \times 4^2 \times 3\frac{1}{2}}{4} = \frac{3.1416 \times 16 \times 3\frac{1}{2}}{4} = 43.98 \text{ cu in.}$$

If the engine has more than one cylinder, the total displacement would be 43.98 times the number of cylinders.

5.9 Compression Ratio. The compression ratio of an engine is a measurement of how much the air-fuel charges are compressed in the engine cylinders. It is calculated by dividing the air volume in one cylinder with the piston at BDC by the air volume with the piston at TDC (Fig. 5.8).

NOTE: The air volume with the piston at TDC is called the clearance volume, since it is the clearance that remains above the piston at TDC.

For example, the engine of one popular automobile has a cylinder volume of 42.35 cu in. at BDC (*A* in Fig. 5.8). It has a clearance volume of 4.45 cu in. (*B* in Fig. 5.8). The compression ratio, therefore, is 42.35 divided by 4.45, or 9.5/1 (that is, 9.5 : 1). In other words, during the compression stroke, the air-fuel mixture is compressed from a volume of 42.35 cu in. to 4.45 cu in., or to 1/9.5 of its original volume.

5.10 Increasing Compression Ratio. In recent years, the compression ratios of automotive and other engines have been repeatedly increased. This increase offers several advantages. The power and economy of an engine increase as the compression ratio goes up (within

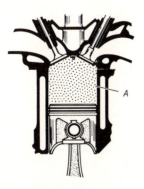

PISTON AT BDC

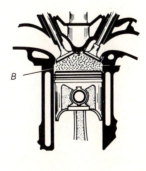

PISTON AT TDC

FIGURE 5.8 Compression ratio is the volume in a cylinder with the piston at BDC divided by its volume with the piston at TDC, or *A* divided by *B*.

limits) without a comparable increase in engine size or weight. In effect, an engine with a higher compression ratio "squeezes" the air-fuel mixture harder (compresses it more). This causes the air-fuel mixture to produce more power on the power stroke. Here is the reason: A higher compression ratio means a higher initial pressure at the end of the compression stroke. This means that, when the power stroke starts, higher combustion pressures will be attained: a harder push will be registered on the piston. The burning gases will also expand to a greater volume during the power stroke. It all adds up to this: There is more push on the piston for a larger part of the power stroke. More power is obtained from each power stroke.

While the compression ratio of small engines has been increased in recent years, there are limits to the compression ratio that can be built into a small engine. It is harder to crank a high-compression engine, and engine parts are stressed more so they must be built stronger. And the high-compression engine is more finicky about the gasoline it will use: it requires a gasoline with a higher antiknock value—that is, a higher octane rating. The high-compression engine has a greater tendency to knock, and it will not tolerate just any fuel. We will get into octane rating of fuel and other characteristics of gasoline in a later chapter.

5.11 Volumetric Efficiency. The amount of air-fuel mixture taken into the cylinder on the intake stroke is a measure of the engine's volumetric efficiency. If the mixture were drawn into the cylinder very slowly, a full measure could get in. But the mixture must pass very rapidly through a series of restricting openings and bends in the carburetor and intake manifold. In addition, the mixture is heated (from engine heat); it therefore expands. The two conditions, rapid movement and heating, reduce the amount of mixture that can get into the cylinder. A full charge of air-fuel mixture cannot enter, because the time is too short and because the air becomes heated.

Volumetric efficiency is the ratio between the amount of air-fuel mixture that actually enters the cylinder and the amount that could enter under ideal conditions. For example, a certain cylinder has an air volume (A in Fig. 5.8) of 47 cu in. If the cylinder were allowed to completely "fill up," it would take in 0.034 oz of air. However, suppose that the engine were running at a fair speed, so that only 0.027 oz of air could enter during each intake stroke. This means that volumetric efficiency would be only about 80 percent (0.027 is 80 percent of 0.034). Actually, 80 percent is a good volumetric efficiency for an engine running at fairly high speed. Volumetric efficiency of some engines may drop to as low as 50 percent at high speeds. This is another way of saying that the cylinders are only "half-filled at the high speeds.

This is one reason why engine speed and output cannot continue to increase indefinitely. At higher speed, the engine has a harder time "breathing," or drawing in air. It is "starved" for air and cannot produce any further increase in power output.

You can now see the scientific reason why the valves in four-cycle engines are timed the way they are. As you remember from Fig. 4.11, a typical valve-timing arrangement starts the intake valve opening at 15° before TDC on the exhaust stroke, and keeps it open until 50° after BDC on the compression stroke. The same valve timing opens the exhaust valve 50° before BDC on the power stroke and keeps it open until 15° after TDC on the intake stroke. This longer valve-open time allows the cylinder to breathe better. That is, it can get rid of the exhaust gases more completely and take in a larger amount of air-fuel mixture so that the power stroke is stronger. More power is produced.

5.12 Brake Horsepower. The horsepower output of engines is measured in terms of brake horsepower (bhp) because a braking device is used to hold the engine speed down while horsepower is measured. When an engine is rated at 300 hp, for example, it is actually brake horsepower that is meant. This is the amount of power the engine can produce at a particular speed at wide-open throttle.

The usual way to rate an engine is with a *dynamometer*. This device has a dynamo, or generator, that is driven by the engine. The amount of electric current the generator produces is a measure of the amount of horsepower the engine is developing.

5.13 Indicated Horsepower. The engine may also be evaluated in terms of ihp (indicated horsepower). Indicated horsepower is based on the power actually developed inside the engine cylinders by the combustion processes. A special indicating device (an oscilloscope) is required to determine ihp. This device measures the pressure continuously throughout the four piston strokes (intake, compression, power, exhaust). A graph of the cylinder pressures taken during a typical test of engine ihp is shown in Fig. 5.9. The four small drawings show the crank, rod, and piston positions as well as directions of motion during the four strokes. Note that the pressure in the cylinder is about atmospheric at the beginning of the intake stroke. Then it falls a little below atmospheric as the delivery of the air-fuel mixture to the cylinder lags behind piston movement (that is, volumetric efficiency is less than 100 percent). When the compression stroke begins, the pressure starts to increase as the piston moves upward in the cylinder. A little before the piston reaches TDC, ignition takes place.

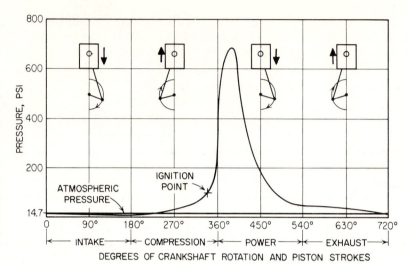

FIGURE 5.9 Pressures in an engine cylinder during the four piston strokes. The four strokes require two crankshaft revolutions (360° each), a total of 720° of rotation. This curve is for a particular engine operating at one definite speed and throttle opening. Changing the speed and throttle opening would change the curve (particularly the power curve.)

Now the air-fuel mixture burns, and pressure goes up very rapidly. It reaches a peak of around 680 psi at about 25° past TDC on the power stroke. Pressure then falls off rapidly as the power stroke continues. But there is still a pressure of around 50 psi at the end of the power stroke. When the exhaust stroke begins, the pressure falls off and drops to about atmospheric at the end of the stroke.

A graph such as the one shown in Fig. 5.9 supplies the information needed to determine ihp. This is because ihp is based on the average pressure during the power stroke minus the average pressures during the other three strokes.

Some of the power developed in the engine cylinders (or ihp) is used in overcoming friction in the engine. Thus, ihp is always greater than bhp (or power delivered by engine).

5.14 Friction Horsepower. Friction losses in an engine are sometimes referred to in terms of fhp (friction horsepower). This expression means the amount of horsepower used up in the engine to overcome friction. Friction horsepower is determined by driving the engine with an electric motor to measure the horsepower required to drive it. During this test, the engine is at operating temperature, but there is no fuel in the carburetor, and the throttle is held wide open. At low speed, friction is relatively low. But as engine speed increases, fhp goes up rapidly. The graph (Fig. 5.10) shows fhp in a typical automobile engine at different speeds. At 1,000 rpm, the fhp is only about

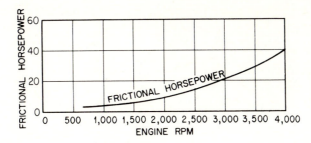

FIGURE 5.10 Friction-horsepower curve, showing the relationship between fhp and engine speed.

4 hp. But, at 2,000 rpm, it is nearly 10 hp. At 3,000 rpm, it is up to 21 hp, and, at 4,000 rpm, it is about 40 hp.

One of the major causes of frictional loss (or fhp) in an engine is piston-ring friction. Under some conditions, the friction of the rings on the cylinder walls accounts for 75 percent of all friction losses in the engine. For example, Fig. 5.10 shows an fhp of 40 hp at 4,000 rpm. It could be that 75 percent, or 30 hp, is due to friction between the rings and cylinder walls. Understanding this fact makes us more fully aware of the difficult job the rings have in the engine. It also points up one advantage of the short-stroke, over-square engine. With a short stroke, the piston rings do not have as far to slide on the cylinder walls, and thus ring friction is reduced. This lowers frictional losses in the engine.

5.15 Relating BHP, IHP, and FHP.   Brake horsepower is the power delivered, ihp is the power developed in the engine, and fhp is the power lost owing to friction. The relationship among the three is

$$bhp = ihp - fhp$$

That is, the horsepower delivered by the engine (bhp) is equal to the horsepower developed (ihp) minus the power lost owing to friction (fhp).

5.16 Engine Torque.   Torque is turning effort. When the piston is moving down on the power stroke, it is applying torque to the engine crankshaft (through the connecting rod). The harder the push on the piston, the greater the torque applied. Thus, the higher the combustion pressures, the greater the amount of torque.

The dynamometer is normally used to check engine torque. Torque can be measured along with horsepower on the dynamometer.

5.17 Brake Horsepower vs. Torque.   The torque that an engine can develop changes with engine speed (see Fig. 5.11). During inter-

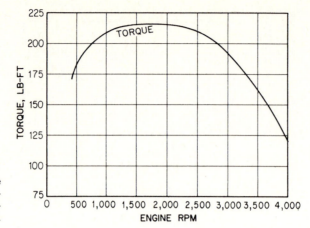

FIGURE 5.11 Torque curve of an engine, showing the relationship between torque and speed.

mediate speeds, volumetric efficiency is high (there is sufficient time for the cylinders to become fairly well "filled up"). This means that, with a fairly full charge of air-fuel mixture, higher combustion pressures will develop. With higher combustion pressures, the engine torque is higher.

But, at higher speed, volumetric efficiency drops off (there is not enough time for the cylinders to become filled up with air-fuel mixture). Since there is less air-fuel mixture to burn, the combustion pressures will not go so high. There will be less push on the pistons, and thus engine torque will be lower. Note in the graph (Fig. 5.11) how the torque drops off as engine speed increases.

The bhp curve of an engine is considerably different from the torque curve. Figure 5.12 is the bhp of the same engine for which the torque curve is shown in Fig. 5.11. It starts low at low speed and increases steadily with speed until a high engine speed is reached. Then, as still higher engine speeds are attained, bhp drops off.

The dropoff of bhp is due not only to reduced torque at higher speed but also to increased fhp at the higher speed. Figure 5.13 compares the curves of these three factors, torque, bhp, and fhp, of an engine.

NOTE: The curves (Figs. 5.10 to 5.13) are for one particular engine only. Different engines have different torque, bhp, and fhp curves. Peaks may be at higher or lower speeds, and the relationships may not be as indicated in the curves shown.

5.18 Engine Efficiency. The term "efficiency" means the relationship between the effort exerted and the results obtained. As applied

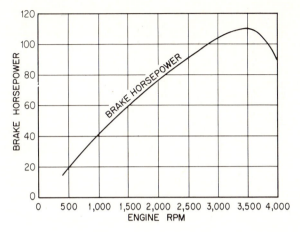

FIGURE 5.12 Curve showing the relationship between bhp and engine speed.

to engines, efficiency is the relationship between the power delivered and the power that could be obtained if the engine operated without any power loss. Engine efficiency can be computed in two ways, as mechanical efficiency and as thermal efficiency.

1. Mechanical Efficiency. This is the relationship between bhp and ihp, or

$$\text{Mechanical efficiency} = \frac{\text{bhp}}{\text{ihp}}$$

2. Thermal Efficiency. "Thermal" means of or pertaining to heat. The thermal efficiency of the engine is the relationship between the power

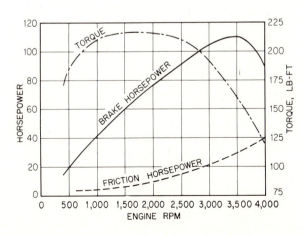

FIGURE 5.13 Torque-bhp-fhp curves of an engine.

output and the energy in the fuel burned to produce this output.

Some of the heat produced by the combustion process is carried away by the engine cooling system. Some of it is lost in the exhaust gases, since they are still hot as they leave the cylinder. These are heat (thermal) losses that reduce the thermal efficiency of the engine. They do not add to the power output of the engine. The remainder of the heat, in causing the gases to expand and produce high pressure, forces the pistons down so that the engine develops power. Because there is a great deal of heat lost during engine operation, thermal efficiencies of the engine may be as low as 20 percent and are seldom higher than 25 percent.

CHECKUP

Work, energy, power, friction—these are all important concepts for anyone dealing with engines. Likewise, the engine expert will know what compression ratio, volumetric efficiency, horsepower, and torque mean and their significance in engine operation and servicing. The following questions will not only give you a chance to check up on how well you understand and remember these fundamentals, but also will help you to remember them better. The act of writing down the answers to the questions will fix the facts more firmly in your mind.

NOTE: Write down your answers in your notebook. Then later you will find your notebook filled with valuable information which you can refer to quickly.

Completing the Sentences: Test 5. The sentences below are not complete. After each sentence there are several words or phrases, only one of which will correctly complete the sentence. Write each sentence in your notebook, selecting the proper word or phrase to complete it correctly.

1. Moving an object against an opposing force is called (*a*) work; (*b*) torque; (*c*) power.
2. The ability to do work is called (*a*) power; (*b*) torque; (*c*) energy.
3. The rate at which work is done is called (*a*) power; (*b*) torque; (*c*) energy.
4. 33,000 ft-lb per minute is called (*a*) power; (*b*) a horsepower; (*c*) torque.
5. Twisting or turning effort is called (*a*) power; (*b*) torque; (*c*) energy.
6. The three classes of friction are (*a*) dry, greasy, viscous; (*b*) dry, liquid, air; (*c*) vapor, liquid, solid.

7. The ratio of the cylinder volume at BDC and the clearance volume is called (*a*) clearance ratio; (*b*) volumetric ratio; (*c*) compression ratio.

8. The ratio between the amount of air-fuel mixture that is taken into the cylinder during normal running and the amount that could enter under ideal conditions is called the (*a*) clearance ratio; (*b*) volumetric efficiency; (*c*) compression ratio.

9. Friction losses in the engine are sometimes referred to as (*a*) bhp; (*b*) ihp; (*c*) fhp.

10. The relationship between bhp, ihp, and fhp is (*a*) ihp = bhp − fhp; (*b*) bhp = ihp + fhp; (*c*) bhp = ihp − fhp.

Written Checkup

In the following, you are asked to write down, in your notebook, the answers to the questions asked or to define certain terms. Writing the answers down will help you to remember them.

1. Define work.
2. Define energy.
3. Define power.
4. What is a horsepower?
5. Name and explain the three classes of friction.
6. What is piston displacement?
7. What is compression ratio? What are the advantages of increasing compression ratio?
8. What is volumetric efficiency? What are the advantages of increasing volumetric efficiency?
9. What is brake horsepower? How is it measured?
10. What is friction horsepower? How is it measured?
11. What is indicated horsepower?
12. How are bhp, ihp, and fhp related?
13. Explain the relationship between engine torque and horsepower.
14. What is mechanical efficiency and how is it calculated?
15. What is thermal efficiency?

Engine Lubrication

6

In this chapter, we will look at the ways in which two-cycle and four-cycle engines are lubricated. We have already mentioned that in the two-cycle engine, the oil is mixed with the gasoline. The oil enters the engine along with the air-fuel mixture and lubricates the engine as we will explain below. In the four-cycle engine, there is a supply of oil in the lower part of the crankcase. When the engine is running, oil from this reserve supply is either pumped or splashed on all moving parts in the engine. In either type of engine, the result is that the engine parts get the lubrication they need so that minimum friction and wear result.

6.1 Lubricating the Engine. When a four-cycle engine, such as the engine used in automobiles, needs oil, it is poured into a pipe or an opening on the side of the engine. From there, the oil runs down into the lower part of the engine — the crankcase. As we will explain later, the crankcase in the four-cycle engine is the source from which the engine lubricating system sends oil to all engine moving parts. In the two-cycle engine, this system would not work. The reason is this: If oil were kept in the crankcase of a two-cycle engine, the incoming air-fuel mixture (which passes through the crankcase) would pick up some of the oil and carry it up into the cylinder. There, the oil would be burned. Soon, all the oil would be used up in this way, and the engine would then fail from lack of oil.

Another lubricating method is called for in the two-cycle engine. This method is to mix a little lubricating oil with the gasoline. The oil therefore enters the crankcase, along with the air-fuel mixture, as a fine mist. Some of the oil mist is carried on up to the engine cylinder where it is burned. But part of it gets on the cylinder wall and engine bearings to provide adequate lubrication. The amount of oil to be added to the gasoline varies with the engine. Some engines require only an ounce or two per gallon of gasoline. Other engines require more. The instructions of the engine manufacturer should always be followed when oil is mixed with gasoline for use in the two-cycle engine.

6.2 Purpose of Lubricating Oil. We normally think of lubricating oil as a substance that minimizes friction and wear between moving parts in a machine. However, the lubricating oil does a number of other things in the engine. The lubricating oil must:

1. Lubricate moving parts to minimize wear
2. Lubricate moving parts to minimize power loss from friction
3. Remove heat from engine parts by acting as a cooling agent
4. Absorb shocks between bearings and other engine parts, thus reducing engine noise and extending engine life
5. Form a good seal between piston rings and cylinder walls
6. Act as a cleaning agent

1 and 2. Minimizing Wear and Power Loss from Friction. Friction has been discussed in some detail. The type of friction encountered in the engine is normally viscous friction, that is, the friction between adjacent moving layers of oil. If the lubricating system does not function properly, sufficient oil will not be supplied to moving parts, and greasy or even dry friction will result between moving surfaces.

This would cause, at the least, considerable power loss, since power would be used in overcoming these types of friction. At most, major damage would occur to engine parts as greasy or dry friction developed. Bearings would wear with extreme rapidity, and the heat resulting from dry or greasy friction would cause bearing disintegration and failure, so that connecting rods and other parts would be broken; also, insufficient lubrication of cylinder walls would cause rapid wear and scoring of walls, rings, and pistons. A properly operating four-cycle engine lubricating system supplies all moving parts with sufficient oil so that only viscous friction is obtained. In the two-cycle engine, adding the proper amount of oil to the gasoline assures adequate engine lubrication.

3. Removing Heat from Engine Parts. In the four-cycle engine, the oil is in rapid circulation throughout the engine lubrication system. All bearings and moving parts are bathed in streams of oil. In addition to providing lubrication in the four-cycle engine, the oil absorbs heat from engine parts and carries it back into the oil pan. The oil pan in turn absorbs heat from the oil, transferring it to the surrounding air. The oil thus acts as a cooling agent.

4. Absorbing Shocks between Bearings and Other Engine Parts. As the piston approaches the end of the compression stroke and the mixture in the cylinder is ignited, pressure in the cylinder suddenly increases many times. A load of as much as a ton is suddenly imposed on the top of the piston as combustion takes place. This sudden increase in pressure causes the piston to thrust down hard through the piston-pin bearing, connecting rod, and connecting-rod bearing. There is always some space, or clearance, between bearings and journals; this space is filled with oil. When the load suddenly increases as described above, the layers of oil between bearings and journals must act as cushions, resisting penetration or "squeezing out," and must continue to interpose a film of oil between the adjacent metal surfaces. In thus absorbing and cushioning the hammerlike effect of the suddenly imposed loads, the oil quiets the engine and reduces wear of parts.

5. Forming a Seal between Piston Rings and Cylinder Walls. Piston rings must form a gastight seal with the cylinder walls, and the lubricating oil that is delivered to the cylinder walls helps the piston rings to accomplish this. The oil film on the cylinder walls compensates for microscopic irregularities in the fit between the rings and walls and fills in any gaps through which gas might escape. The oil film also provides lubrication of the rings, so that they can move easily in the ring grooves and on the cylinder walls.

6. Acting as a Cleaning Agent. The oil, as it circulates through the engine (four-cycle type), tends to wash off and carry away dirt, particles of carbon, and other foreign matter. As the oil picks up this material, it carries it back to the crankcase. There, larger particles drop to the bottom of the oil pan.

6.3 Properties of Oil. A satisfactory engine lubricating oil must have certain characteristics, or properties. It must have proper viscosity (body and fluidity) and must resist oxidation, carbon formation, corrosion, rust, extreme pressures, and foaming. Also, it must act as a good cleaning agent, must pour at low temperatures, and must have good viscosity at extremes of high and low temperature.

Any mineral oil, by itself, does not have all these properties. Lubricating-oil manufacturers therefore put a number of additives into the oil during the manufacturing process. An oil for severe service may have additives, as follows:

1. Usually a viscosity-index improver
2. Pour-point depressants

3.   Oxidation inhibitors
4.   Corrosion inhibitors
5.   Rust inhibitors
6.   Foam inhibitors
7.   Detergent dispersants
8.   Extreme-pressure agents

These are discussed below.

1. Viscosity (Body and Fluidity).   Primarily, viscosity is the most important characteristic of lubricating oil. Viscosity refers to the tendency of oil to resist flowing. In a bearing and journal, layers of oil adhere to the bearing and journal surfaces. These layers must move, or slip, with respect to each other, and the viscosity of the oil determines the ease with which this slipping can take place. Viscosity may be divided for discussion into two parts, body and fluidity. Body has to do with the resistance to oil-film puncture, or penetration, during the application of heavy loads. When the power stroke begins, for example, bearing loads sharply increase. Oil body prevents the load from squeezing out the film of oil between the journal and the bearing. This property cushions shock loads, helps maintain a good seal between piston rings and cylinder walls, and maintains an adequate oil film on all bearing surfaces under load.

Fluidity has to do with the ease with which the oil flows through oil lines and spreads over bearing surfaces. In some respects, fluidity and body are opposing characteristics, since the more fluid an oil is, the less body it has. The oil used in any particular engine must have sufficient body to perform as explained in the previous paragraph and yet must have sufficient fluidity to flow freely through all oil lines and spread effectively over all bearing surfaces. Late types of engines have more closely fitted bearings with smaller clearances and consequently require oils of greater fluidity that will flow readily into the spaces between bearings and journals.

Temperature influences viscosity. Increasing temperature reduces viscosity. That is, it causes oil to lose body and gain fluidity. Decreasing temperature causes oil viscosity to increase. The oil gains body and loses fluidity. Since engine temperatures range several hundred degrees from cold-weather starting to operating temperature, a lubricating oil must have adequate fluidity at low temperatures so that it will flow. At the same time, it must have sufficient body for high-temperature operation.

2. Viscosity Ratings.   Viscosity of oil is determined by use of a viscosimeter, a device that determines the length of time required for a definite amount of oil to flow through an opening of a definite diameter. Temperature is taken into consideration during this test, since high temperature decreases viscosity, while low temperature increases viscosity. In referring to viscosity, the lower numbers refer to oils of lower viscosity (thinner). The Society of Automotive Engineers (SAE) rates oil viscosity in two different ways, for winter and for other than winter. Winter-grade oils are tested at 0° and 210° F. There are three grades, SAE5W, SAE10W, and SAE20W, the W indicating winter grade. For other than winter use, the grades are SAE20, SAE30, SAE40, and SAE50, all without the "W" suffix. Some oils have multiple ratings, which means they are equivalent, in viscosity, to several single-rating oils. An SAE10W-30 oil, for example, is comparable to SAE10W, SAE20W, and SAE30 oils.

3. Viscosity Index.   When oil is cold, it is thicker and runs more slowly than when it is hot. In other words, it becomes more viscous when it is cooled. On the other hand, it becomes less viscous when it is heated. In normal engine operation, we do not have to be too concerned about this change of oil viscosity with changing temperature. We recognize that the engine is harder to start at low temperature because the oil is thicker, or more viscous. But until the engine is cooled to many degrees below zero, we do not have to take any special steps to start it.

Some oils change viscosity a great deal with temperature change. Other oils show a much smaller change of viscosity with temperature change. In order to have an accurate measure of how much any particular oil will change in viscosity with temperature change, the viscosity-index scale was adopted. Originally, the scale ran from 0 to 100. The higher the number, the less the oil viscosity changes with temperature changes. Thus, an oil with a VI (viscosity index) of 100 will change less in viscosity with temperature changes than an oil with a VI of 10. In recent years, special VI-improving additives have been developed which step up viscosity indexes to as much as 300. Such an oil shows relatively little change in viscosity from very low to relatively high temperature.

You could especially appreciate the significance of VI if you were operating automotive equipment in a very cold climate (say, in northern Alaska). You would have to start engines at temperatures as much as 60° below zero (92° below freezing). But, once started, the engines would soon reach operating temperatures that heat the oil to several hundred degrees. If you could select an oil of a relatively high VI, then it would be fluid enough to permit starting but would not thin out (or lose viscosity) so much that lubricating effectiveness would be lost. On the other hand, an oil with a low VI would probably be so thick at low temperatures that it might actually prevent starting. But if you could start, it might then thin out too much as it warmed up.

Oil companies make sure that their oils have a sufficiently high VI to operate satisfactorily in the variations of temperatures they will meet. Also, they supply oil with multiple-viscosity ratings. For example, an oil may be designated SAE10W-30, which indicates that it is comparable to SAE10W, SAE20W, and SAE30 oils.

4. Pour-point Depressants. At low temperatures, some oils become so thick that they will not pour at all. Certain additives can be put into oil which will depress, or lower, the temperature point at which the oil will become too thick to flow. Such additives keep the oil fluid at low temperatures for adequate engine lubrication during cold-weather starting and initial operation.

5. Resistance to Carbon Formation. Cylinder walls, pistons, and rings operate at temperatures of several hundred degrees. This temperature, acting on the oil films covering walls, rings, and pistons, tends to cause the oil to break down or burn so that carbon is produced. Carbon formation can cause poor engine performance and damage to the engine. Carbon may pack in around the piston rings, causing them to stick in the ring grooves. This prevents proper piston-ring operation, so that blowby, poor compression, excessive oil consumption, and scoring of cylinder walls may result. Carbon may build up on the piston head and in the cylinder head. This fouls spark plugs, excessively increases compression so that knocking occurs, and reduces engine performance. Carbon may form on the underside of the piston to such an extent that heat transfer will be hindered and the piston will overheat. In four-cycle engines, pieces of carbon may break off and drop into the oil pan, where they may be picked up by the lubrication system. They could then clog oil channels and lines so that the flow of lubricating oil to engine parts would be dangerously reduced. A good lubricating oil must be sufficiently resistant to the heat and operating conditions in the engine to exhibit a minimum amount of carbon formation.

6. Oxidation Inhibitors. When oil is heated to fairly high temperatures and then agitated so that considerable air is mixed with it, the oxygen in the air tends to combine with oil, oxidizing it. Since this is the treatment that engine oil undergoes in four-cycle engines (that is, it is heated and agitated with or sprayed into the air in the crankcase), some oil oxidation is bound to occur. A slight amount of oxidation will do no particular harm, but if oxidation becomes excessive, serious troubles may occur in the engine. As the oil is oxidized, it breaks down to form various harmful substances. Some of the products of oil oxida-ion coat engine parts with an extremely sticky, tarlike material. This material may clog oil channels and tend to restrict the action of piston rings and valves. A somewhat different form of oil oxidation coats engine parts with a varnish-like substance that has a similar dam-aging effect on the engine. Even if these substances do not form, oil oxidation may produce corrosive materials in the oil that will corrode bearings and other surfaces, causing bearing failures and damage to other parts. Oil chemists and refineries control the refining processes and may add certain chemicals known as oxidation inhibitors so that engine lubricating oils resist oxidation.

7. Corrosion and Rust Inhibitors. At high temperatures, acids may form in the oil which can corrode engine parts, especially bearings. Corrosion inhibitors are added to the oil to inhibit this corrosion. Also, rust inhibitors are added. These displace water from metal surfaces so that oil coats them. Also, they have an alkaline reaction to neutralize combustion acids.

8. Foaming Resistance. The churning action in the engine crankcase in four-cycle engines also tends to cause the engine oil to foam, just as an egg beater causes an egg white to form a frothy foam. As the oil foams up, it tends to overflow, or to be lost through the crankcase ventilator. In addition, the foaming oil is not able to provide normal lubrication of bearings and other moving parts. Foaming oil in hy-draulic valve lifters will cause them to function poorly, work noisily, wear rapidly, and possibly break. To prevent foaming, antifoaming additives are mixed with the oil.

9. Detergent Dispersants. Despite the filters and screens at the carburetor and crankcase ventilator, dirt does get into the engine. In addition, as the engine runs, the combustion processes leave deposits of carbon on piston rings, valves, and other parts. Also, some oil oxidation may take place, resulting in still other deposits. Then, too, metal wear in the engine puts particles of metal into the oil. As a result of these various conditions, deposits tend to build up on and in engine parts. The deposits gradually reduce the performance of the engine and speed up wear of parts. To prevent or slow down the formation of these deposits, some engine oils contain a detergent additive.

A detergent oil is very useful in a four-cycle engine for the fol-lowing reason. The detergent acts much like ordinary hand soap.

When you wash your hands with soap, the soap surrounds the particles of dirt on your hands, causing them to become detached so that the water can rinse them away. In a similar manner, the detergent in the oil loosens and detaches the deposits of carbon, gum, and dirt. The oil then carries the loosened material away. The larger particles drop to the bottom of the crankcase, but smaller particles tend to remain suspended in the oil. These impurities, or contaminants, are flushed out when the oil is changed.

To prevent the particles from clotting, and to keep them in a finely divided state, a dispersant is added to the oil. Without the dispersant, the particles would tend to collect and form large particles. These large particles might then block the oil filter and reduce its effectiveness. They could also build up in oil passages and plug them, thus depriving bearing and other engine parts of oil. The dispersant prevents this and thus greatly increases the amount of contaminants the oil can carry and still function effectively.

Lubricating-oil manufacturers now place more emphasis on the dispersant qualities of the additive than on its detergent qualities. If the contaminants can be kept suspended in the oil as small particles, they will not deposit on engine parts and there is less need for detergent action.

A detergent oil is of no value in a two-cycle engine because the oil does not circulate and thus cannot carry away deposits, as it does in the four-cycle engine.

10. Extreme-Pressure Resistance. The modern automotive engine subjects the lubricating oil to very high pressures, not only in the bearings, but also in the valve train. Modern valve trains have heavy valve springs and high-lift cams. This means that the valves must move farther against heavier spring opposition. To prevent the oil from squeezing out, extreme-pressure additives are put into the oil. They react chemically with metal surfaces to form very strong, slippery films which may be only a molecule or so thick. Thus, they supplement the oil by providing protection during moments of extreme pressure.

6.4 Water-Sludge Formation. Water sludge is a thick, creamy, black substance that often forms in the crankcase of four-cycle engines. It clogs oil screens and oil lines, preventing normal circulation of lubricating oil to engine parts. This can result in engine failure from oil starvation.

1. How Sludge Forms. Water collects in the crankcase in two ways. First, water is formed as a product of combustion. Second, the crankcase ventilating system carries air, with moisture in it, through the crankcase. If the engine parts are cold, the water condenses and drops into the crankcase. There, it is churned up with the lubricating oil by the action of the crankshaft. The crankshaft acts much like a giant eggbeater and whips the oil and water into the thick, black, mayonnaiselike "goo" known as water sludge. The black color comes from dirt and carbon.

2. Why Sludge Forms. If a four-cycle engine is operated an hour or more each time it is started, the water that collects in the crankcase while the engine is cold quickly evaporates. The crankcase ventilating system then removes the water vapor. Thus, no sludge will form. However, if the engine is operated when cold most of the time, then sludge will form.

3. Preventing Sludge. To prevent sludge, the engine must be operated long enough, after being started, for it to heat up and get rid of the water in the crankcase. If this is impractical, then the oil must be changed frequently. Naturally, during cold weather, it takes longer for the engine to warm up. Thus, in cold weather, the operating time must be still longer or oil must be changed still more frequently, to prevent water accumulation and sludge formation.

6.5 Service Ratings of Oil. We have already mentioned that lubricating oil is rated as to its viscosity by number. Lubricating oil is also rated in another way, by what is called service designation. That is, it is rated according to the type of service for which it is best suited. There are five service ratings for gasoline engines: SA, SB, SC, SD, and SE. There are four service ratings for diesel engines: CA, CB, CC, and CD. The oils differ in their characteristics and in the additives they contain.

1. SA Oil. Oil for utility gasoline and diesel engines operating under mild conditions so that the protection provided by additives is not required. This oil may have pour-point and foam depressants.

2. SB Oil. This oil is for service in gasoline engines operated under such mild conditions that only minimum protection provided by additives is required. Oils designed for this service have been used since the 1930s and provide only antiscuff capability and resistance to oil oxidation and bearing corrosion.

3. SC Oil. This oil is for service typical of gasoline engines in the 1964–1967 models of passenger cars and trucks and is intended primarily for use in passenger cars. This oil provides control of high- and low-temperature engine deposits, wear, rust, and corrosion.

4. SD Oil. This oil is for service typical of gasoline engines in passenger cars and trucks beginning with 1968 models. This oil provides more protection from high- and low-temperature engine deposits, wear, rust, and corrosion than the SC oils.

5. SE Oil. This oil is for service typical of gasoline engines in passenger cars and some trucks beginning with 1972 (and some 1971) models. This oil provides more protection against oil oxidation, high-temperature engine deposits, rust, and corrosion than do oils with the SC and SD ratings.

Diesel-engine oils must have different characteristics than oils for gasoline engines, and the designations CA, CB, CC, and CD indicates oils for increasingly severe diesel-engine operation. For example, CA oil is for light-duty service while CD oil is for severe-duty service typical of high speed, high output diesel engines.

All car manufacturers recommend the use of a high-detergent engine oil. A high-detergent oil is designated by HD on the can. Here, HD means "high detergency" as well as "heavy duty."

> CAUTION. **Do not confuse viscosity and service ratings of oil. Some people think that a high-viscosity oil is a "heavy-duty" oil. This is not necessarily so. Viscosity ratings refer to the thickness of the oil; thickness is not a measure of heavy-duty quality. Remember that there are two ratings, viscosity and service. Thus, an SAE 10 oil can be an SC, SD, or SE oil. Likewise, an oil of any other viscosity rating can have any one of the service ratings.**

6.6 Oil Changes. From the day that fresh oil is put into the engine crankcase (four-cycle engines), it begins to lose its effectiveness as an engine lubricant. This gradual loss of effectiveness is largely due to the accumulation of various contaminating substances. For instance, water sludge may accumulate, as already noted. In addition, during engine operation, carbon tends to form in the combustion chamber. Some of this carbon gets into the oil. Gum, acids, and certain lacquer-like substances may also be left by the combustion of the fuel or may be produced in the oil itself by the high engine temperatures. In addition, the air that enters the engine (in the air-fuel mixture) carries with it a certain amount of dust. Even though the air filter is operating efficiently, it will not remove all the dust. Then, too, the engine releases fine metal particles as it wears. All these substances tend to circulate with the oil. As the operating hours pile up, the oil accumulates more and more of these contaminants. Even though the engine has an oil filter, some of these contaminants will remain in the oil. Finally, after many hours of operation, the oil will be so loaded with contaminants that it is not safe to use. Unless it is drained and clean oil put in, engine wear will increase rapidly.

The oil should be changed at recommended intervals. In automotive engines, this might be every two months or 6,000 miles, whichever comes first. For small four-cycle engines, many manufacturers recommend oil changes every 25 hr of operation. Remember that engines in power mowers and other such equipment often operate in unusually dusty conditions, and the oil must be changed frequently to prevent heavy accumulations of dirt in the oil that could cause rapid engine wear. Always follow the engine manufacturer's recommendations on oil changes. Usually, you will find these recommendations printed on the nameplate that is attached to the engine or to the equipment in which the engine is installed.

6.7 Types of Four-Cycle Engine Lubricating Systems. We have already mentioned that the two-cycle engine does not have a lubricating system, as such. These engines depend for lubrication on the oil that is mixed with the gasoline. In four-cycle engines, however, there is an oil reservoir in the crankcase and the lubricating system has the job of circulating this oil through the engine. The oil lubricates all engine parts, provides some cooling of engine parts, helps keep engine parts clean, and so on, as we have already explained.

Small four-cycle engines use several methods to provide lubrication for the moving parts. The simplest means of lubrication is to splash the oil about so that all parts are drenched (Fig. 6.1). The splashing is accomplished by attaching a dipper to the connecting-rod cap (Fig. 6.2). The oil in the crankcase is splashed every time the dipper reaches into it on the downstroke of the piston. Splash may also be accomplished by the use of an oil slinger which is rotated by the camshaft gear (Fig. 6.3).

A cam operated barrel-type pump is used in certain small engines to provide a pressure-type lubrication system (Figs. 6.4 to 6.6). A small gear-type pump operated by the camshaft gear is sometimes used on small vertical-shaft engines (Fig. 6.7).

Figure 6.8 shows the lubrication system on a small multicylinder air-cooled engine, which uses spray nozzles to direct oil to the moving parts.

The engine lubricating system forces or splashes oil on all moving surfaces and thus provides ample lubrication to keep friction low. But the oil does many other jobs, as we have previously explained.

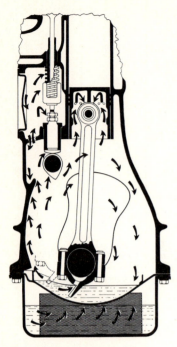

FIGURE 6.1 Splash lubricating system on an L-head engine. The dipper on the connecting rod splashes oil, as shown by the arrows, every time the piston passes through BDC.

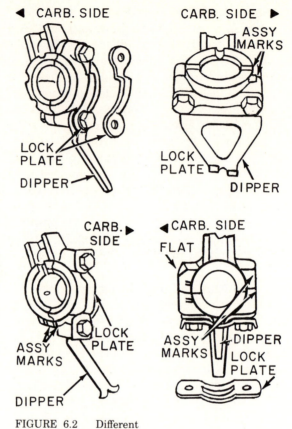

FIGURE 6.2 Different ways in which the dipper is mounted on the connecting-rod cap.

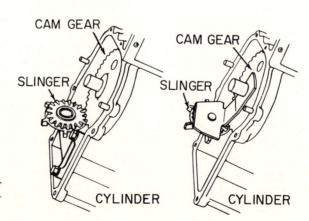

FIGURE 6.3 Oil slinger is rotated by the camshaft gear.

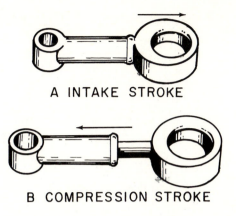

A INTAKE STROKE

B COMPRESSION STROKE

C CUTAWAY VIEW

FIGURE 6.4   Barrel-type lubrication pump used on the Lawson vertical-shaft engine.

FIGURE 6.6   Location of the barrel-type lubrication pump and the oil passages in the engine. (*Tecumseh Products Company.*)

FIGURE 6.5   The plunger-type oil pump is assembled on an eccentric on the camshaft. (*Tecumseh Products Company.*)

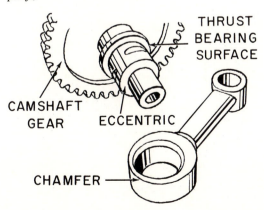

THRUST BEARING SURFACE

CAMSHAFT GEAR

ECCENTRIC

CHAMFER

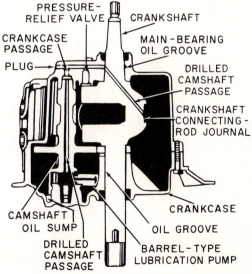

PRESSURE-RELIEF VALVE

CRANKSHAFT

CRANKCASE PASSAGE

MAIN-BEARING OIL GROOVE

PLUG

DRILLED CAMSHAFT PASSAGE

CRANKSHAFT CONNECTING-ROD JOURNAL

CAMSHAFT OIL SUMP

OIL GROOVE

CRANKCASE

DRILLED CAMSHAFT PASSAGE

BARREL-TYPE LUBRICATION PUMP

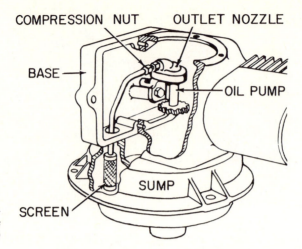

FIGURE 6.7 Gear-driven oil pump. (*Briggs and Stratton Corporation.*)

Automotive engines (the four-cycle type) use a pressure-feed oiling system; that is, the oil flows under pressure to all moving engine parts (Fig. 6.9). Several quarts of oil are held as a reservoir in the oil pan at the bottom of the crankcase. An oil pump, driven by a gear

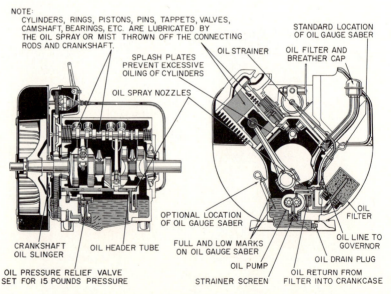

FIGURE 6.8 Multiple-cylinder air-cooled engine lubrication system. (*Wisconsin Motor Corporation.*)

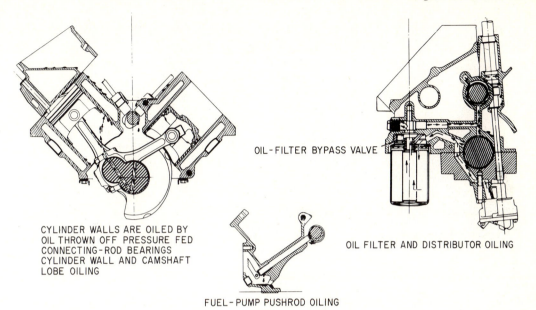

CYLINDER WALLS ARE OILED BY
OIL THROWN OFF PRESSURE FED
CONNECTING-ROD BEARINGS
CYLINDER WALL AND CAMSHAFT
LOBE OILING

OIL-FILTER BYPASS VALVE

OIL FILTER AND DISTRIBUTOR OILING

FUEL-PUMP PUSHROD OILING

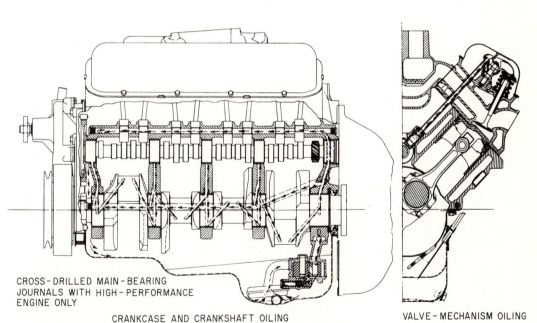

CROSS-DRILLED MAIN-BEARING
JOURNALS WITH HIGH-PERFORMANCE
ENGINE ONLY

CRANKCASE AND CRANKSHAFT OILING

VALVE-MECHANISM OILING

FIGURE 6.9 Lubrication system of a V-8, overhead-valve engine. (*Chevrolet Motor Division of General Motors Corporation.*)

off the camshaft, pumps oil from the oil pump through holes drilled in the cylinder block and metal tubes to the various engine parts. A gear-type oil pump is shown in Fig. 6.10. As the gears rotate past the oil inlet of the pump, the spaces between the teeth become filled with oil. Then, when the teeth mesh, the oil is forced out from between the teeth and through the oil-pump outlet.

Another type of oil pump is shown in Fig. 6.11. This pump has two rotors, as shown. The inner rotor is mounted on the oil-pump shaft and turns with it. As it turns, it carries the outer rotor around. The spaces between the rotor lobes become filled with oil as they move past the oil-pump inlet. Then, when the lobes of the inner rotor move into the spaces between the lobes in the outer rotor, the oil is squeezed out through the oil-pump outlet.

Some of the oil lubricates the crankshaft, connecting-rod, and piston-pin bearings. Here is the path it follows to do this. First, the oil is forced through oil holes drilled in the cylinder block to holes in the main bearings which support the crankshaft. This oil fills grooves cut into the bearings. From there, oil spreads out to cover and lubricate the bearings and the crankshaft journals turning in them.

Part of the oil entering the oil grooves in the main bearings feeds through holes drilled in the crankshaft from the main bearings to the connecting-rod bearings. These holes are drilled at angles to permit the oil to flow to the rod bearings. The oil flows out on the rod bearings to cover and lubricate the bearings and the crankpins turning in them.

Some of the oil flowing to the connecting-rod bearings has a further job to do, and that is to lubricate the piston-pin bearing. In

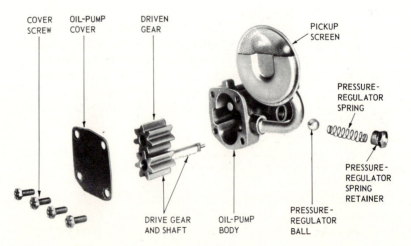

FIGURE 6.10 Disassembled view of a gear-type oil pump. (*Pontiac Motor Division of General Motors Corporation.*)

COVER SCREW   OIL-PUMP COVER   DRIVEN GEAR   PICKUP SCREEN   PRESSURE-REGULATOR SPRING   PRESSURE-REGULATOR SPRING RETAINER   DRIVE GEAR AND SHAFT   OIL-PUMP BODY   PRESSURE-REGULATOR BALL

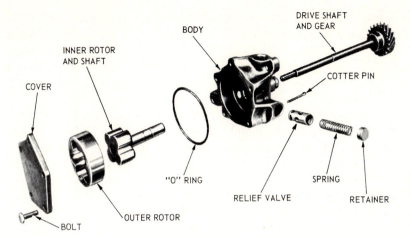

FIGURE 6.11 Disassembled view of a rotor-type oil pump. (*Dodge Division of Chrysler Motors Corporation.*)

some engines each connecting rod has an oil hole drilled in it from the connecting-rod-bearing surface to the piston-pin bearing. Every time the hole in the connecting-rod bearing registers (lines up) with the hole in the crankpin—which is once a revolution—a spurt of oil flows up through the hole in the connecting rod and onto the piston-pin bearing.

The oil from all these bearings is thrown off and flows back down to the crankcase. Some of it sprays out onto the cylinder walls and lubricates the walls, pistons, and rings.

CHECKUP

The chapter you have just completed discusses lubricating oil, its composition, its purpose, and its various ratings. It also explains why the engine parts must be kept lubricated and how the two-cycle and four-cycle engines are lubricated. The following questions will not only give you a chance to check up on how well you understand and remember these fundamentals, but also will help you to remember them better. The act of writing down the answers to the questions will fix the facts more firmly in your mind.

NOTE: **Write down your answers in your notebook. Then later you will find your notebook filled with valuable information which you can refer to quickly.**

Completing the Sentences: Test 6. The sentences below are not complete. After each sentence there are several words or phrases, only one of which will correctly complete the sentence. Write each sentence in your notebook, selecting the proper word or phrase to complete it correctly.

1. The two-cycle engine is lubricated (*a*) from oil in the crankcase; (*b*) from oil in the gasoline; (*c*) from an oil pump.
2. In addition to lubricating, the engine oil must (*a*) oxidize, carbonize, and burn; (*b*) absorb shocks, enter as mist, clean; (*c*) absorb shocks, seal, and clean.
3. The tendency of oil to resist flowing is called its (*a*) rating; (*b*) fluidity; (*c*) viscosity.
4. The three winter grades of oil are SAE5W, SAE10W, and (*a*) SAE20W; (*b*) SAE30W; (*c*) SAE15W.
5. A common method of lubricating moving parts in small four-cycle engines is to have a dipper on the (*a*) camshaft; (*b*)crankshaft; (*c*) connecting rod.
6. The reason the crankcase in the two-cycle engine cannot be used as an oil reservoir is that the crankcase also serves as (*a*) a pre-combustion chamber; (*b*) an exhaust chamber; (*c*) passage for the incoming air-fuel mixture.
7. The three service ratings of oil for engines are (*a*) MS, MM, and ML; (*b*) MM, DD, SS; (*c*) MA, MB, and MC.
8. A typical recommendation for changing oil in a small engine would be to change it after operating the engine (*a*) 250 hr; (*b*) 6,000 miles; (*c*) 25 hr.

Written Checkup

In the following, you are asked to write down, in your notebook, the answers to the questions asked or to define certain terms. Writing the answers down will help you to remember them.

1. Explain why the oil must be added to the gasoline in the two-cycle engine.
2. Why can't the crankcase in the two-cycle engine be used as an oil reservoir?
3. Make a list of the various jobs the lubricating oil does in a four-cycle engine.
4. Explain what oil viscosity is and why it is important to the proper lubrication of the engine.
5. What are the viscosity ratings for winter-grade oil?
6. What are the service ratings of oil for spark-ignition engines, and what do they mean?

7. Describe the oil pumps used in four-cycle-engine lubricating systems.
8. Describe the path the lubricating oil takes in a typical engine as it leaves the crankcase, goes through the engine, and then returns to the crankcase.

Ignition Systems

7

In this chapter, we are going to describe the ignition systems used in small engines. First, however, let us review the story of electricity. We know that a flow of electric current in a wire is actually a movement of electrons. To see where electrons come from, and how they are made to move in wires, we will take another look at atoms. You will remember that we discussed atoms in Chap. 2.

7.1 Atoms and Electrons.   Every element in the world is composed of atoms. Atoms are exceedingly small, far too small to be seen by even the most powerful microscope. But scientists have studied their actions and have constructed a picture of the atom.

Atoms, small as they are, are made up of still smaller units. For instance, an atom of the gas *hydrogen* consists of two parts, the center, or nucleus, of the atom and a small particle which whirls around the nucleus (Fig. 7.1). The center, or nuclear, particle is called the *proton,* and it has a charge of positive electricity. The outer particle is called an electron, and it has a charge of negative electricity.

7.2 Size of Atoms, Electrons, and Protons.   At 32° F and atmospheric pressure, a cubic inch of hydrogen gas would contain about 880 billion billion atoms. If we were to expand this 1-in. cube until it was large enough to contain the earth—that is, to a cube 8,000 miles on a side—a single atom in this tremendously enlarged cube would measure about 10 in. in diameter.

The atom is tiny, but the particles that make up the atom are far smaller. For example, it would take something like a hundred thousand billion electrons, side by side, to measure an inch. It is impossible, of course, to see these particles with a microscope.

7.3 Attraction and Repulsion.   In the hydrogen atom, a single electron whirls around a single proton, just as the earth orbits the sun. The electron is kept in its circular orbit by a combination of

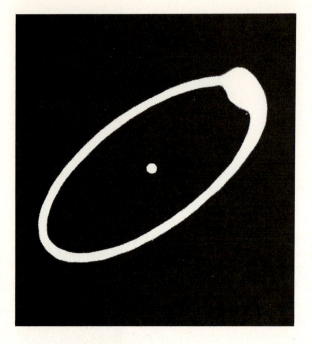

FIGURE 7.1 The hydrogen atom consists of two particles; a proton with a positive electric charge and an electron with a negative electric charge.

forces. One force is the attraction that positive and negative electrical charges have for each other (protons are positive, electrons negative). Unlike electrical charges attract each other.

An opposing force is *centrifugal*. Centrifugal force attempts to get the electron to fly in a straight path. Compare this force with the force acting on a ball swung on a rubber band around your hand (Fig. 7.2). As the ball is swung, centrifugal force tries to make it fly away in a straight path. But the rubber band exerts an attractive force and keeps the ball moving in a circle.

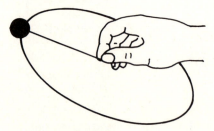

FIGURE 7.2 The electron in a hydrogen atom circles the proton like a ball on a rubber band swung in a circle around the hand.

7.4 Electric Current. Atoms often lose their electrons. When these "lost" electrons gather in one area, that area has an extra number of electrons and therefore a charge of negative electricity. If there is a path along which these electrons can flow, they will move away from this concentration to any place in which there are atoms that are missing some of their electrons. This movement is called a *current of electricity* or *electric current*.

One reason the electron movement takes place is that electrons repel each other. In other words, like electrical charges repel. Negative repels negative, positive repels positive. Remember, too, that unlike charges attract. That is, positive attracts negative.

Thus, the electron movement is caused by a combination of the two conditions: like charges repelling and unlike charges attracting. For example, when we connect an electric circuit, or electron path, to a battery, electrons will flow from the negative terminal of the battery through the electron path to the positive terminal. There is a concentration of electrons at the negative battery terminal and a shortage of electrons at the positive battery terminal. The battery, by chemical action, produces this unbalanced condition. Therefore, if a path is provided, electrons will flow in an attempt to restore a balance of electrical charges.

7.5 Conductors. A conductor is an electron path through which electrons, that is, electricity, can flow easily. Copper is a good conductor. The reason for this is that copper atoms can very easily lose electrons. There are great numbers of electrons moving about in all directions among the atoms of copper (Fig. 7.3).

If the copper wire is not connected to a source of electrons, such as a battery, the electrons move about in all directions in the wire. But when the wire is connected between the terminals of a battery, then electrons pour into the wire from the negative terminal. Electrons begin to move in the same direction all along the wire, moving from the negative terminal and pouring out of the wire into the positive terminal. The negative charge at the negative terminal pushes elec-

FIGURE 7.3 In a copper wire, many free electrons move from atom to atom.

trons into the wire. The positive charge at the positive terminal attracts, or "pulls" the electrons out of the other end of the wire. A flow of electrons, or an electric current, results.

7.6 Insulators. Insulators are substances that are composed of atoms which hold on to their electrons. Thus, in such substances, there are few, if any, free electrons moving around. Therefore, when an insulating material is placed around a negative terminal of a battery, for example, the electrons on the negative terminal cannot push into the insulating material. The insulating material stops the flow of current. Insulators are very important and wires carrying electric current must be covered with insulation to prevent the current from leaking off through the supports holding the wire in place. Current-carrying parts in electrical devices, such as generators or motors, are either supported so that they do not touch other parts, or are covered with insulating material such as mica, rubber, bakelite, fiber, or porcelain.

7.7 Electromagnetism. Electric batteries concentrate electrons at one terminal and take them from the other terminal. They produce this unbalanced condition by chemical means. An imbalance of electrical charge can also be produced by mechanical means, that is, by moving a magnet past a conductor. A magnet will attract bits of iron such as tacks, paper clips, and hairpins. This is called *magnetic attraction,* or *magnetism* (Fig. 7.4). Magnetism also has an effect on electrons. For instance, if a wire is moved past a magnet (Fig. 7.5), the free electrons in the wire are moved in one direction. In other words, they stop moving about in random directions and start moving in the same direction along the wire. This action is the basis of the ignition system for the two-cycle engine and also of the generator or alternator in the automotive electric system.

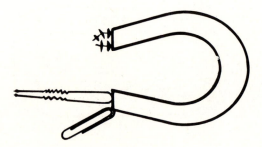

FIGURE 7.4   A magnet will attract objects made of iron.

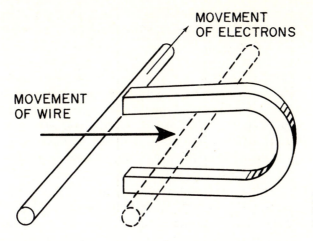

MOVEMENT
OF ELECTRONS

MOVEMENT
OF WIRE

FIGURE 7.5   Current
flows in a conductor when
it moves past a magnet.

The magnet produces this effect by means of invisible magnetic lines of force that stretch in the air between the two poles of the magnet (Fig. 7.6). When these lines of force are cut by a conductor — that is, by a wire having free electrons in it — the magnetic action gives the electrons a push so they will all move in the same direction. Magnetism can produce current; and, conversely, current can produce magnetism.

Conductors are surrounded by magnetic lines of force whenever current is flowing through them. The magnetism produced by electric current is called *electromagnetism*. Automotive generators or alternators, regulators, ignition coils, etc., contain conductors which are used to produce electromagnetism.

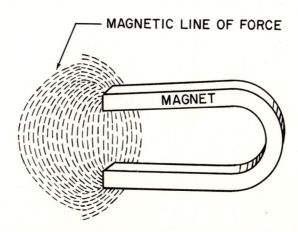

MAGNETIC LINE OF FORCE

MAGNET

FIGURE 7.6   A magnet
has lines of force stretching between its poles.

7.8 Small-Engine Magneto Ignition Systems. Some small engines use battery ignition systems that are somewhat similar to the systems used in automobiles. Most, however, use magneto ignition systems. We will describe the magneto ignition system first. The magneto ignition system in the typical one-cylinder, two-cycle engine is built into the engine. Usually, it is located at one end of the crankshaft. The principle of operation is simple. A series of magnets are whirled past a coil of wire. The magnets are mounted on the engine flywheel. See Fig. 7.7. When magnetic lines of force move through a conductor, voltage is induced in the conductor. If the conductor is in a closed circuit, then current will flow.

Figure 7.8 is a wiring diagram of the primary circuit. It includes the coil of wire, the primary winding, past which the magnets move. Also included in this circuit is a pair of breaker points. One of these points is on a lever, or arm. The other is stationary. One end of the arm rests on a plunger which rides on a cam on the crankshaft. This cam is round except for a high spot. When the crankshaft and cam rotate, the breaker points remain closed until the high spot passes under the plunger. The plunger is raised so that the contact points are separated.

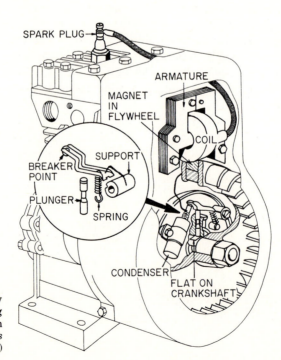

FIGURE 7.7 Cutaway view of engine, showing location and construction of the magneto. (*Briggs and Stratton Corporation.*)

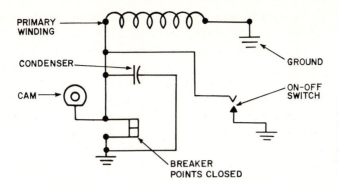

FIGURE 7.8 Wiring diagram for magneto ignition with current flowing through the primary circuit. (*Lawn Boy Division, Outboard Marine Corporation.*)

Now, let us see how these actions can produce an electric spark. When the engine is running, the magnets are whirling past the coil primary winding. Voltage is induced and current flows when the breaker points are closed. This current causes a strong magnetic field to build up around the winding. When the high spot on the cam comes around under the plunger and the breaker points separate, the current stops flowing. Now the magnetic field rapidly collapses.

The capacitor—or condenser as it is also called—aids this rapid collapse of the magnetic field. It contains two long strips of metal foil insulated from each other. When the points start to separate, the current would continue to flow, causing a momentary arc between the points, if it were not for the capacitor. But for a moment the capacitor provides a place for this current to flow. It acts somewhat like a check spring and brings the current to a quick stop. This produces the rapid magnetic-field collapse.

Surrounding the primary winding is a secondary winding made of many thousands of turns of a fine wire. See Fig. 7.9. The magnetic

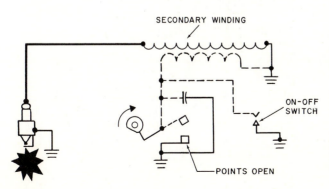

FIGURE 7.9 Wiring diagram of magneto ignition system with breaker points open. Current has stopped flowing in the primary circuit, and a high-voltage surge has been induced in the secondary circuit to produce a spark at the spark plug gap in the cylinder. (*Lawn Boy Division, Outboard Marine Corporation.*)

field from the primary winding, in collapsing, moves rapidly through the secondary winding. Since this is a movement of a magnetic field through a conductor, a voltage is induced in the secondary winding. And, since there are many thousands of turns of wire in the secondary winding, a high voltage is induced.

The spark plug, shown in Fig. 7.10, is connected to the two ends of the secondary winding. One end is connected through the metal of the engine (called ground), and the other through a rubber-covered wire (called the high-tension lead). The voltage in the secondary winding quickly goes up high enough to cause a powerful spark to occur at the gap between the two spark plug electrodes. One of these electrodes is connected to the metal shell of the plug which is screwed into the cylinder head of the engine. The other is insulated by a porcelain shell in which it is centered. The porcelain is breakable, just like glass, and this is the reason that the center electrode must never

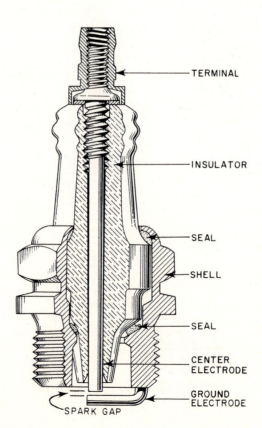

FIGURE 7.10 Spark plug partly cut away to show construction.

TERMINAL

INSULATOR

SEAL

SHELL

SEAL

CENTER ELECTRODE

GROUND ELECTRODE

SPARK GAP

be bent when the spark plug gap is adjusted. Only the outer electrode should be bent. If the center electrode is bent, it probably will break the porcelain and ruin the plug. This is also the reason that the plug must be removed and installed with care; improper handling will also break the porcelain and ruin the plug. We will have more to say about spark plugs on a later page.

Figure 7.11 shows, in end view, what happens before and after the contact points separate. Figure 7.12 is a top view of a typical magneto. Notice the arrangement of the coil, the contact points, and the condenser. The magnets are curved so that, as the flywheel rotates, they pass close to the coil.

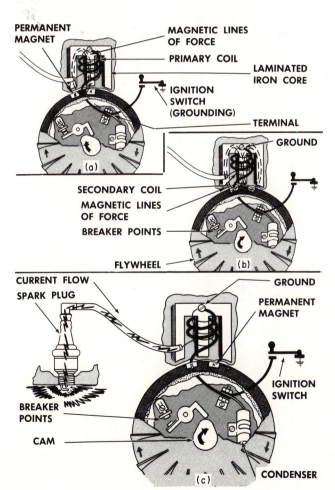

FIGURE 7.11 Principles of the flywheel magneto ignition system. (*a*) Magnetic lines of force are built up around a coil of wire by the movement of a permanent magnet past the iron core on which the coil is wound. (*b*) As the permanent magnet passes the iron core and the contact points open, the lines of force collapse, thus producing a high voltage in the secondary winding of the coil. (*c*) The high voltage produces a spark at the spark plug gap.

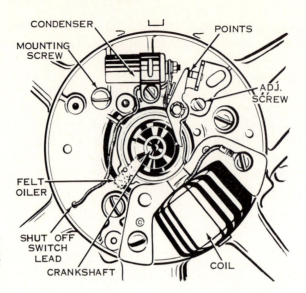

FIGURE 7.12   Top view of the magneto. (*Lawn Boy Division, Outboard Marine Corporation.*)

An ON-OFF switch is used on many systems to turn the engine off. See Fig. 7.9. When this switch is flipped so it is closed, it grounds the contact-point end of the primary winding. Now, current continues to flow in the primary winding, and opening the points does not interrupt it. As a result, no sparks occur and the engine stops. The engine can also be stopped by a grounding blade located near the spark plug which can be bent by hand—or foot—to ground the insulated terminal of the plug, as shown in Fig. 7.13. When this happens, the high-voltage surges flow through the blade, and no spark occurs.

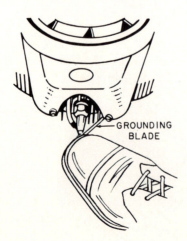

FIGURE 7.13 Grounding blade near the spark plug used to stop the engine.

7.9 External Magneto.　Some engines have an externally mounted magneto, as shown in Fig. 7.14. The magneto rotor is driven through an impulse coupling which will be explained later. As the rotor spins, it produces a magnetic field in the laminated iron frame on which the primary and secondary coils are wound. Each half turn of the magnetic rotor causes a complete reversal of the magnetic field in the laminated iron frame. This, in turn, causes magnetic lines of force to build up and collapse through the primary and secondary windings. Thus, a flow of current is induced in the primary winding all the time that the contact points are closed.

When the current flow is at its greatest, the breaker points are opened by the cam on the end of the rotor shaft. This stops the flow of current, and the magnetic lines of force therefore collapse very rapidly.

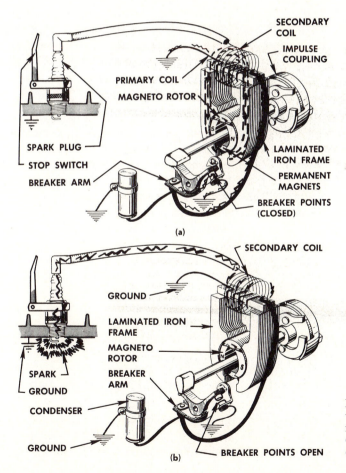

FIGURE 7.14 Schematic view of an external-type magneto ignition system. (*a*) Breaker points closed. (*b*) Breaker points open.

The rapid movement of the lines of force through both the primary and secondary windings produces a high voltage in the secondary winding. This voltage is high enough to produce a strong spark at the spark plug gap. The condenser does the same job here as in the other magneto ignition system previously discussed.

The impulse coupling through which the rotor shaft is driven is included to improve starting. It does two things. First, it retards the ignition timing for better starting during cranking. Second, it flips the magneto rotor at the proper moment so that it spins very rapidly for a part turn and thus produces a stronger spark. The faster the magnetic field from the magneto rotor moves through the laminated iron frame, the stronger the magnetic field induced in the iron frame and the higher the voltage in the secondary winding goes. The impulse coupling produces this action through a delayed spring action. That is, during cranking, spring tension builds up during a part turn of the coupling and then releases to spin the rotor ahead. The rotor thus turns part way, stops momentarily until spring tension builds up again, and is once more flipped ahead. After the engine starts, the impulse coupling unlocks due to centrifugal action so that it does not function. Now, the rotor turns steadily in time with the engine.

7.10 Solid-state Ignition.   This system does not use contact points and thus less maintenance is required. Instead, it uses a series of diode rectifiers and a transistorized switching device somewhat similar to those used in automotive transistorized ignition systems. (See Fig. 7.15.) The rotating magnet and primary winding on an iron frame are similar to the arrangement on magnetos previously discussed. Rotation of the flywheel carries the magnets past the iron frame, and this induces an alternating current in the primary winding, as shown in Fig. 7.15*a*. Now, a small current is induced in the trigger coil and this current flows to the transistorized switch. The current is sufficient to cause the transistorized switch to close, in effect, so that the charge in the condenser can flow through the primary winding of the ignition coil. The charge is sufficient to cause the ignition coil to produce a high voltage in its secondary winding with the result that the spark occurs at the spark plug, and the compressed air-fuel charge in the cylinder is ignited.

A schematic layout of a typical solid-state ignition system is shown in Fig. 7.16. This system is somewhat different from the one illustrated in Fig. 7.15. The system in Fig. 7.16 uses a battery as the source of electric power rather than magnets on the flywheel and a primary winding next to the flywheel. In operation, current from the battery flows through the trigger module and the primary winding of

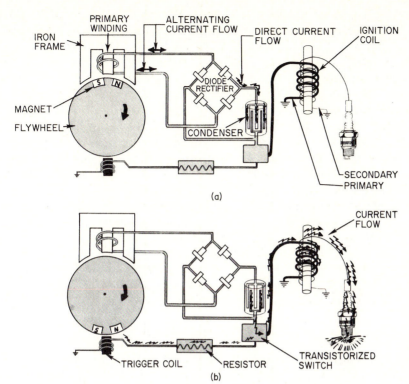

(a)

(b)

FIGURE 7.15 Schematic view of a solid-state ignition system. The transistorized switch is essentially a switching device that, when triggered, allows a flow of current to pass through it. Or, when not triggered, it will prevent a flow of current.

the ignition coil. Then, as the flywheel rotates to bring the metal projection around past the trigger module, the electronic circuit halts this flow of current. The magnetic field in the ignition coil primary winding collapses and a high-voltage surge is produced in the ignition-coil secondary winding. The spark plug fires and the power stroke takes place.

The system shown in Fig. 7.16 includes an alternator, as you can see. The alternator produces current which flows through the rectifier-regulator to the battery. Thus, the battery is kept charged. We will cover alternators in a later chapter.

Another system combines an alternator and a magneto. The windings for this system are shown in Fig. 7.17. The magneto works in the same manner as other flywheel magnetos previously described. The alternator, which in this system is combined with the magneto, will be described later.

7.11 Battery Ignition Systems. The battery ignition system differs in two basic respects from the magneto ignition system. First, the

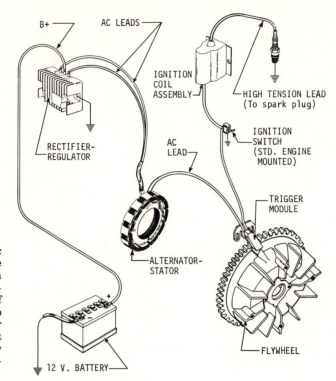

FIGURE 7.16 Schematic diagram of a solid-state ignition system for a small engine using a battery as the source of power. This system also includes an alternator which produces current to keep the battery charged. (*Kohler Company.*)

current is supplied by a battery or generator, just as in automotive ignition systems. Second, the ignition switch in the battery ignition system must be *closed* for the system to work. Also, the battery ignition system has an ignition coil. This coil contains two separate windings, a primary winding of a few hundred turns of relatively heavy wire, and a secondary winding of many thousands of turns of very fine wire. Otherwise, the two systems are very similar. Operation of the battery ignition system is shown in Figs. 7.18 and 7.19. Rotation of a cam causes the contact points to close and open. When the contact points are closed, the primary winding of the ignition coil is connected to the battery. This allows current to flow through the primary winding and build up a strong magnetic field. Then, when the cam rotates so the lobe on the cam opens the contact points, the current stops flowing in the primary winding. The magnetic field collapses and this causes the secondary winding to produce a very strong pulse of high-voltage current. The high voltage causes a spark at the spark plug which ignites the compressed air-fuel mixture in the engine cylinder. Aiding in the quick collapse of the magnetic field is the condenser. It protects

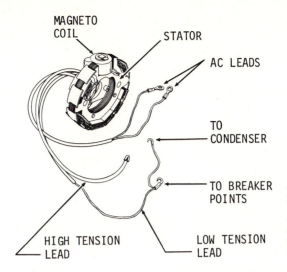

FIGURE 7.17 The combined alternator stator and magneto coil for an ignition system that also includes an alternator. (*Kohler Company.*)

the contact points by providing a momentary place for the current to flow as the contact points begin to separate. Otherwise, the current would try to continue to flow and would arc across the contact points, burning them. The condenser prevents this, and at the same time brings the current flow to a quick stop. This hastens the collapse of

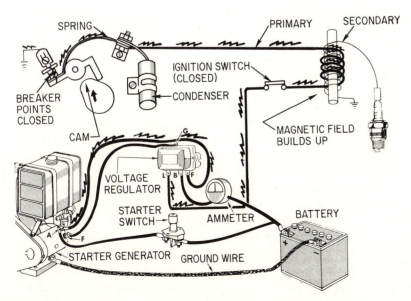

FIGURE 7.18 Schematic view of a battery ignition system with the breaker points closed and current flowing from the battery through the primary winding of the ignition coil.

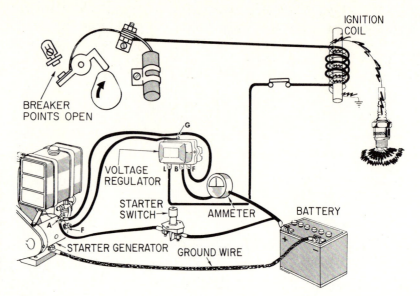

FIGURE 7.19 Schematic view of a battery ignition system with the contact points open. The collapsing magnetic field in the ignition coil produces a high voltage in the coil secondary winding, and this causes a spark to occur at the spark plug.

the magnetic field and thereby increases the high voltage in the secondary winding. We will have more to say about this in following paragraphs when we discuss ignition systems for automotive-type engines.

7.12 Automotive-type Ignition Systems. Ignition systems for automobiles are somewhat more complicated than the system illustrated in Figs. 7.18 and 7.19. They have to supply sparks for several cylinders and they have to do this with great regularity and with great speed. For example, an ignition system for an eight-cylinder engine that is pushing a car along at highway speed must produce something like 15,000 sparks per minute.

The battery ignition system for any engine with two or more cylinders consists of an electric storage battery, a switch, an ignition distributor, ignition coil, spark plugs, and wiring. Figure 7.20 shows, in schematic form, a typical ignition system. This is for an eight-cylinder engine, but only one of the eight spark plugs required is shown for the sake of simplicity.

7.13 Ignition Distributor. Ignition distributors for multicylinder engines are shown in Figs. 7.21 and 7.22. The distributor has two jobs. First, it closes and opens the circuit between the battery and ignition coil. With the circuit closed, current flows through the ignition coil and builds up a magnetic field. When the circuit opens, the magnetic field collapses and a high-voltage surge of current is produced by the coil. As its second job, the distributor distributes each high-voltage surge to the correct spark plug at the correct instant (it does this through the distributor rotor and cap and the wiring).

FIGURE 7.20 Typical automotive ignition system. It consists of the battery (source of power), ignition switch, ignition coil (shown schematically), distributor (shown in top view with cap removed and placed below it), spark plugs (one shown in sectional view), and wiring. The coil is shown schematically with magnetic lines of force indicated. (*Delco-Remy Division, General Motors Corporation.*)

FIGURE 7.21 Partly disassembled distributor. (*Delco-Remy Division, General Motors Corporation.*)

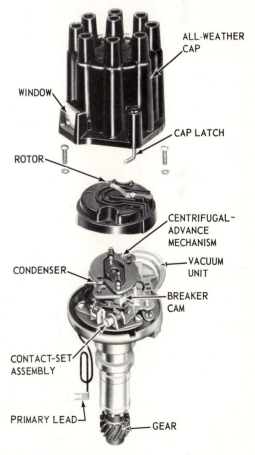

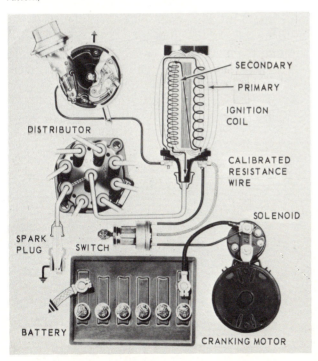

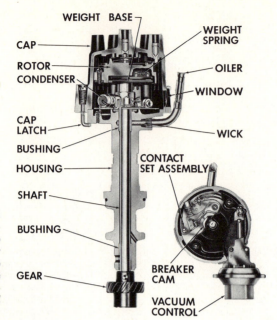

FIGURE 7.22 Sectional and top views of an ignition distributor. In the top view, the cap and rotor have been removed so that the breaker plate may be seen. (*Delco-Remy Division, General Motors Corporation.*)

Figure 7.23 shows an ignition system schematically. The distributor rotor rotates and connects the ignition coil and the different spark plugs in the engine. It makes these connections in the regular firing order, that is, in the order in which the cylinders fire. Thus, every

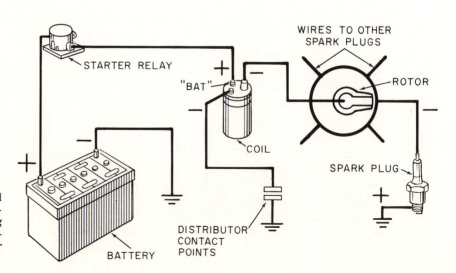

FIGURE 7.23 Simplified wiring diagram of an ignition system, showing position of rotor in distributor. (*Ford Motor Company.*)

time the distributor contact points open, the coil produces a high-voltage surge. At this instant the rotor is in position to direct this high-voltage surge to the spark plug in the cylinder that is ready to fire. In this cylinder, the piston is approaching TDC (top dead center) on the compression stroke.

The distributor consists of a housing, drive shaft with breaker cam and advance mechanism, a breaker plate with contact points, a rotor, and a cap. The shaft is driven by the engine camshaft through a pair of spiral gears, and it rotates at one-half crankshaft speed.

Rotation of the shaft and breaker cam causes the distributor contact points to open and close. There are the same number of lobes, or high spots, on the cam as there are cylinders in the engine. There is one lobe for each cylinder. The contact points thus close and open once for each cylinder with every breaker-cam rotation. Since the breaker cam turns once for every two crankshaft revolutions, one high-voltage surge is thus produced (by closing and opening of the contacts) for each cylinder every two crankshaft revolutions.

The rotor is mounted on top of the breaker cam so they rotate together. As the rotor turns, a metal spring and segment connects the center terminal of the distributor cap with each outside terminal in turn. This connects the ignition coil (which produces the high-voltage surges) with each outside terminal, one after another. Since the outside terminals are connected to the spark plugs, the coil therefore becomes connected to the spark plugs. Thus, as each high-voltage surge is produced in the coil, it is directed through the distributor rotor, cap, and wiring, to the cylinder which is ready to fire.

7.14 Spark Plug. The spark plug (Fig. 7.10) consists of a metal shell, and a porcelain insulator with an electrode extending through its center. Thus, the electrode is insulated from the shell. The shell has a short electrode attached to one side and bent in so there is a small gap between it and the center electrode. This is the gap which the high-voltage surge jumps to produce the ignition spark. The shell has threads so that it can be screwed into a tapped hole in the cylinder head.

One important characteristic of spark plugs that service technicians should know is spark plug heat range. This refers to how hot the tip end of the spark plug gets in operation. Figure 7.24 illustrates what heat range means. A plug that runs hot has a long tip which requires the heat to travel farther from the hot electrodes to the cooler cylinder head. A plug that runs cold has a short tip and thus a much shorter path for the heat to travel. Spark plugs must be matched to engines so that they will operate at the proper temperature. If a plug

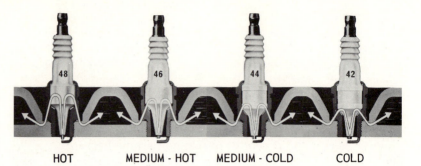

FIGURE 7.24 Heat range in spark plugs. The longer the heat path (indicated by arrows) the hotter the plug runs. (*AC Spark Plug Division of General Motors Corporation.*)

HOT   MEDIUM - HOT   MEDIUM - COLD   COLD

gets too hot, the electrodes will be burned away very quickly so that a new plug will be required. Also, the electrodes may become so hot that they cause preignition. That is, the compressed air-fuel mixture is ignited too soon by the hot electrodes. This can cause severe knocking and broken engine parts. If a plug does not get hot enough, a sooty carbon deposit will form on the tip. This is because the plug does not get hot enough to burn away the carbon. Carbon deposits can cause the plug to foul out. That is, the spark will leak across the carbon instead of jumping the gap. As a result, the engine will not fire and power will be lost or the engine will completely stop. The correct plug is one that stays hot enough to prevent carbon formation and yet will not get so hot as to burn the electrodes away rapidly and result in short plug life.

7.15 Ignition Coil. The ignition coil (Fig. 7.25) steps up the 6 or 12 volts of the battery to the high voltage required to jump the spark plug gap. There are two circuits through the coil, the primary circuit and the secondary circuit. The primary circuit is made up of a few hundred turns of heavy wire. In the ignition system, it is connected to the battery through the ignition switch and distributor contact points. The secondary circuit is made up of many thousands of turns of a very fine wire. It becomes connected to the different spark plugs through the distributor cap and rotor. The high-voltage surge is actually induced in the secondary winding of the ignition coil.

When the distributor contact points close, current flows through the primary winding of the ignition coil. This produces a magnetic field in the ignition coil. Then, when the distributor contact points open, current stops flowing. The magnetic field therefore collapses. The ignition capacitor, connected across the contact points, causes the magnetic field to collapse very rapidly.

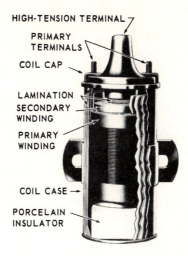

HIGH-TENSION TERMINAL

PRIMARY
TERMINALS

COIL CAP

LAMINATION
SECONDARY
WINDING

PRIMARY
WINDING

COIL CASE →

PORCELAIN
INSULATOR

FIGURE 7.25 Ignition coil with case cut away to show how primary winding is wound around outside of secondary winding. (*Delco-Remy Division, General Motors Corporation.*)

The ignition capacitor, or condenser as it is also called, is made up of two long strips of metal foil, separated by strips of insulation and rolled into a winding (Fig. 7.26). The winding is assembled into a metal case with one of the foil strips connected to the case. The other foil strip is connected to the insulated lead coming out of the end of the condenser. To understand how the condenser operates, look at what would happen if there were no condenser in the ignition system. In

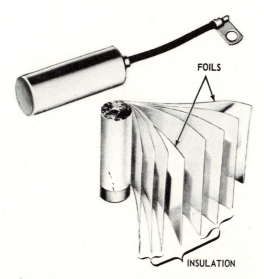

FOILS

INSULATION

FIGURE 7.26 A condenser assembled and with the winding partly unwound.

this case, when the contact points started to separate, the current would momentarily continue to flow and would form an arc between the separating points. This would not only burn the contact points, but also much of the electrical energy in the ignition coil would be used up in the arc. There would not be enough left over to produce the high-voltage surge; no spark would occur at the spark plugs.

The condenser, however, momentarily provides a place for the current to flow. It acts as a sort of check spring and brings the current to a quick stop. Therefore, the magnetic field resulting from the current flow quickly collapses.

The rapid-magnetic-field collapse produces high voltage in the secondary winding. This high voltage then surges through the distributor rotor and cap to the spark plug in the cylinder that is ready to fire. A spark is produced at the spark plug gap, and the compressed air-fuel mixture is ignited.

7.16 Ignition-Coil Resistor.   Many modern cars use 12-volt systems and have a resistor or resistance wire connected in series with the primary circuit when the engine is operating (Fig. 7.20). The resistor acts as a current control, thus protecting the distributor contact points from excessive current. During cranking, the resistor is shorted out. Full battery voltage is thus imposed on the coil, and ignition performance during cranking is therefore improved.

7.17 Centrifugal Advance Mechanism.   When the engine is idling, the spark is timed to occur just before the piston reaches top dead center on the compression stroke. At higher speeds the spark should occur earlier. This gives the mixture time to burn and deliver its power to the piston. To provide this advance, a centrifugal device is built into the distributor (Fig. 7.27). It has a pair of weights that are thrown out against spring tension as the engine speed increases. This movement is transmitted through a toggle arrangement to the breaker cam, causing it to advance or turn ahead with respect to the drive shaft. This, in turn, causes the cam to open and close the contact points earlier in the compression stroke (at higher speeds). Since the rotor, too, is advanced (it is mounted on the breaker cam), it comes into position earlier in the cycle also. The timing of the spark to the cylinder consequently varies from no advance at low speed to full advance at high speed when the weights have reached the outer limit of their travel. Maximum advance may be as much as 45° of crankshaft rotation before the piston reaches TDC. It varies considerably with dif-

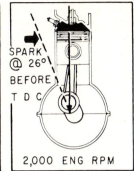

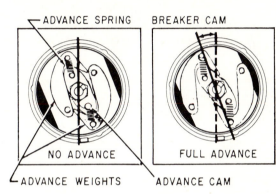

FIGURE 7.27 Centrifugal advance mechanism in no-advance and full-advance positions. In the typical example shown, the ignition is timed at 8° before top dead center on idle. There is no centrifugal advance at 1,000 engine rpm, but there is 26° total advance (18° centrifugal plus 8° due to original timing) at 2,000 engine rpm. (*Delco-Remy Division, General Motors Corporation.*)

ferent makes of engines. The toggle shape and springs are arranged to give the correct advance at every speed to provide best engine performance.

7.18 Vacuum Advance. When the throttle is only partly open, a partial vacuum develops in the intake manifold. This means that less air-fuel mixture enters the cylinder, and it will be less highly compressed. The mixture will burn more slowly. To get full power from the charge, the spark should be somewhat advanced. This would give the mixture ample time to burn and deliver its power to the piston. One type of vacuum advance mechanism is shown in Fig. 7.28. The device has a spring-loaded diaphragm connected through linkage to the distributor breaker plate. The spring-loaded side of the diaphragm is airtight and is connected through a tube to an opening in the carburetor air horn (Fig. 7.29). When the throttle is opened, it moves past this opening so that intake-manifold vacuum can act on the diaphragm. The diaphragm is therefore forced to move against the

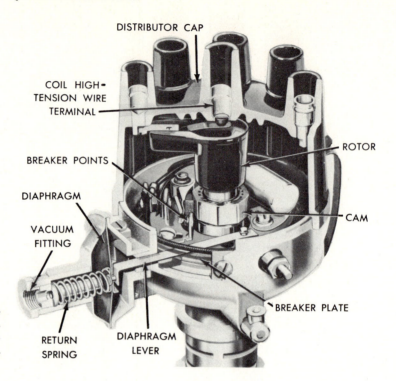

DISTRIBUTOR CAP

COIL HIGH-
TENSION WIRE
TERMINAL

ROTOR

BREAKER POINTS

DIAPHRAGM

CAM

VACUUM
FITTING

BREAKER PLATE

RETURN
SPRING

DIAPHRAGM
LEVER

FIGURE 7.28 Cutaway view of distributor showing construction of vacuum advance mechanism. (*Ford Motor Company.*)

spring. This movement turns the breaker plate in the distributor housing, and the breaker plate with the contact points is moved ahead. The cam, as it rotates, now opens the contact points earlier in the compression stroke, and the spark appears earlier in this stroke. As the throttle is opened wider, there is less vacuum in the intake manifold and less vacuum advance. The mixture, being more highly compressed, burns more rapidly so that less vacuum advance is required.

7.19 Combination Advance. At any particular speed there will be a certain advance due to the centrifugal mechanism plus possibly some additional advance due to the vacuum advance mechanism. Figure 7.30 illustrates this. At 40 mph (miles per hour), for example, the centrifugal advance mechanism provides 15° spark advance on this particular application. The vacuum advance mechanism can supply up to 15° additional advance under part-throttle operation. However, if the engine is operated at wide-open throttle, this added vacuum advance will not be obtained. The advance will thus vary between the straight line (centrifugal) and the curved line (centrifugal plus possible vacuum advance) as the throttle position is changed.

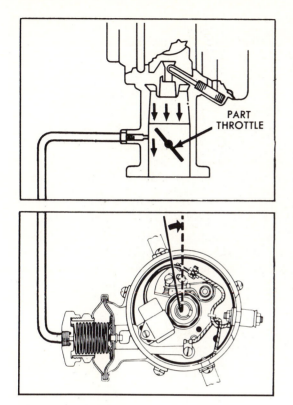

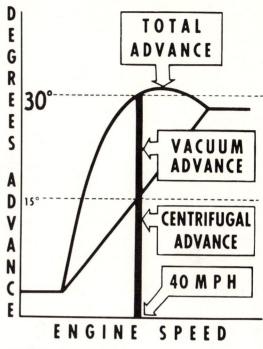

FIGURE 7.30 Centrifugal and vacuum advance curves for one application.

FIGURE 7.29 Operation of vacuum advance mechanism. When the throttle valve swings past the opening, a vacuum is admitted to the vacuum advance mechanism on the distributor and the breaker plate is rotated to advance the spark. (*Delco-Remy Division, General Motors Corporation.*)

CHECKUP

The chapter you have just completed describes ignition systems for small engines. You learned the fundamentals of electricity that apply to small-engine ignition systems, the principles of magneto and battery ignition systems, how these systems work, and the theories behind them. The following questions will not only give you a chance to check up on how well you understand and remember these fundamentals, but also will help you to remember them better. The act of writing down the answers to the questions will fix the facts more firmly in your mind.

NOTE: Write down your answers in your notebook. Then later you will find your notebook filled with valuable information which you can refer to quickly.

Completing the Sentences: Test 7. The sentences below are not complete. After each sentence there are several words or phrases, only one of which will correctly complete the sentence. Write each sentence in your notebook, selecting the proper word or phrase to complete it correctly.

1. Electric current is a flow of (*a*) atoms; (*b*) protons; (*c*) electrons.
2. Substances made up of atoms that hold on to their electrons make good (*a*) conductors; (*b*) insulators; (*c*) metals.
3. Magnetism can produce current, and conversely (*a*) magnetism can produce electrons; (*b*) current can produce magnetism; (*c*) magnets can produce electromagnetism.
4. The principle of magnetos is simple; magnets are moved past (*a*) a coil of wire; (*b*) the flywheel; (*c*) the armature.
5. When the magneto breaker points open, the primary-winding magnetic field (*a*) builds up; (*b*) collapses; (*c*) reacts through the points.
6. The capacitor is connected (*a*) across the primary winding; (*b*) across the secondary winding; (*c*) across the breaker points.
7. When the magnetic field of the primary winding collapses, it induces (*a*) high voltage in the secondary; (*b*) discharging of the capacitor; (*c*) voltage reversal in the primary winding.
8. Two types of magneto used on small engines are (*a*) flywheel and internal; (*b*) flywheel and external; (*c*) flywheel and battery.
9. In the solid-state magneto ignition system, the alternating current is changed to direct current by (*a*) diodes; (*b*) transistors; (*c*) a condenser.
10. While the ignition switch in the magneto ignition system must be open for the system to work, in the battery ignition system, the switch must (*a*) also be open; (*b*) be closed; (*c*) be closed, then opened.

Written Checkup

In the following, you are asked to write down, in your notebook, the answers to the questions asked or to define certain terms. Writing the answers down will help you to remember them.

1. Explain how electrons are set into motion to produce electric current.
2. What is the basic difference between a conductor and an insulator?
3. Explain the basic principle of the flywheel magneto.
4. What is the purpose of the capacitor?
5. Describe an external magneto and explain the purpose of the impulse coupling.
6. Describe the solid-state ignition system and explain how it works.
7. Describe an ignition distributor and explain how it operates.
8. What is heat range in spark plugs? Why is this important?
9. What is the purpose of the centrifugal advance mechanism in the distributor?
10. What is the purpose of the vacuum advance mechanism in the distributor?

Fuel Systems for Small Engines

8

The function of the fuel system is to supply the engine with a mixture of air and gasoline vapor. The mixture must have the proper proportions of air and gasoline for good engine operation. If the mixture does not have enough gasoline vapor (mixture too lean), or if the mixture has too much gasoline vapor (mixture too rich), the engine will not run properly. Also, to start a cold engine, the mixture must be enriched, that is, it must have a higher proportion of gasoline vapor in it. Before we discuss fuel systems for small engines, let us take a look at gasoline.

8.1 Gasoline. Gasoline is a hydrocarbon. That is, it is made up of hydrogen and carbon compounds. These compounds split into hydrogen and carbon atoms when gasoline burns. These atoms then unite with oxygen atoms in the air to form hydrogen oxide (which is water) and carbon dioxide. Gasoline is made from crude oil, or petroleum. The gasoline is extracted from the crude oil by a refining process. Actually, there is more to gasoline than hydrocarbon compounds. The refiners put in small quantities of certain substances, called additives, to improve engine performance. Among these are chemicals to prevent formation of gum, antirust agents to protect metals the gasoline comes in contact with, anti-icers to prevent carburetor icing, detergents to keep the carburetor clean, and most important perhaps, antiknock compounds to prevent engine knocking.

8.2 Antiknock Value. During normal combustion in the engine cylinder, an ever-increasing pressure occurs. Under some conditions, the last part of the compressed air-fuel mixture will explode to produce a sudden and sharp pressure increase. This can cause a rapping or knocking noise that sounds almost as though the piston head had been struck a hard hammer blow. This can be very damaging to the engine, and it can cause rapid engine wear and even breakage of engine parts, especially the piston.

We described, in Sec. 5.9, what compression ratio is and noted that increasing the compression ratio of an engine tends to increase its economy and power. There are problems with this, however, because, with a higher compression ratio, the air-fuel mixture is compressed more. With higher pressure to start with at the instant of ignition, the more ready the mixture is to explode. This means that the higher compression ratios can cause knocking. The gasoline refiners, however, have an answer to this. They make gasolines that are more resistant to knock. One way to do this is to add tetraethyl lead, commonly called "Ethyl," to the gasoline. A few spoonfuls of this compound added to a gallon of gasoline makes the gasoline far more resistant to knocking. The use of this compound, and of other refining operations which improved the antiknock value of the gasoline, has made it possible to produce the high-compression engines used in today's automobiles.

Antiknock value of gasoline is referred to as octane number. A high-octane fuel, say of 98 octane, is much more resistant to knock than an 80 octane fuel. You should always use the gasoline that has the proper octane rating for the engine you are using.

While this is very important if you are driving a late-model automobile, it is less important for small engines. Small engines do not have the high compressions you find in automobile engines. If you use the gasoline the engine manufacturer recommends, you will not have any trouble with knocking—provided everything else is okay. However, if the ignition is out of time so the spark occurs too early in the compression stroke, you can have knock. As you know, in normal operation the spark occurs as the piston nears TDC on the compression stroke. This gives the mixture enough time to ignite and burn while the piston goes over TDC and starts down on the power stroke. But if the spark occurs too early, then the pressure will go up excessively before the piston reaches TDC. This pressure increase will cause knocking and, as we have said, can seriously damage the engine.

There is another possible cause of knock in a small engine. If a great deal of carbon has accumulated in the combustion chamber, the compression ratio will be too high. You will remember that the compression ratio is the ratio of the cylinder volume with the piston at BDC divided by the cylinder volume when the piston is at TDC. If you reduce the TDC cylinder volume, you raise the compression ratio. Accumulation of carbon does this. So an engine that will run normally when it is clean can start to knock when the combustion chamber becomes clogged with carbon. Carbon will accumulate from poor combustion or from excessive amounts of oil getting up into the combustion chamber where it burns and leaves a carbon deposit.

8.3 A Typical Small-Engine Fuel System. We will now take a look at a typical fuel system used on a two-cycle engine in a power mower. This fuel system is shown in Fig. 8.1. A similar system is shown in Fig. 8.11. On a later page, we will look at other types of fuel systems used on small engines. The engine with which this fuel system is used is described and illustrated in Chap. 3. The fuel system consists of the fuel tank with filter, and the carburetor with air filter.

When the engine is running, air passes through the air filter on its way to the carburetor. In the carburetor, it picks up a charge of gasoline and the mixture passes the reed valve as it enters the crankcase of the engine. This action has already been discussed in Chap. 3.

The carburetor itself is shown in sectional view in Fig. 8.2. Essential parts include the air filter, float bowl, choke valve, throttle valve, and the fuel nozzle with adjustment needle.

FIGURE 8.1 Fuel system for two-cycle engine. (*Lawn Boy Division, Outboard Marine Corporation.*)

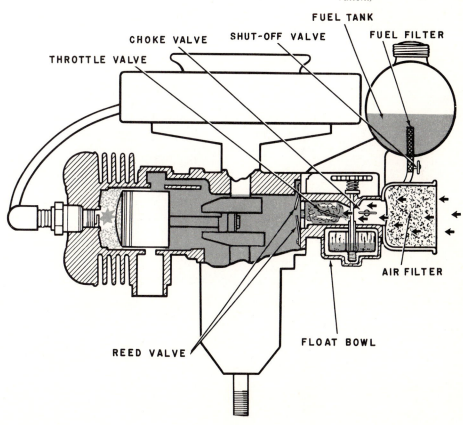

THROTTLE VALVE

CHOKE VALVE

SHUT-OFF VALVE

FUEL TANK

FUEL FILTER

AIR FILTER

FLOAT BOWL

REED VALVE

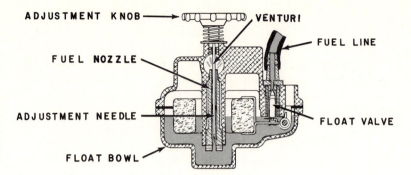

FIGURE 8.2 Sectional view of a carburetor for a two-cycle engine used on a lawn mower. (*Lawn Boy Division, Outboard Marine Corporation.*)

8.4 Fuel Tank. The fuel tank is made of sheet metal. It has a filler cap which can be removed to add gasoline. The filler cap has a small hole for air to enter the tank as gasoline is used. The fuel filter at the tank outlet filters out dirt that might have entered the tank, preventing it from entering the carburetor where it could clog fuel circuits and stop the engine.

8.5 Air Filter. The engine shown in Fig. 8.1 uses a metal mesh air filter. The ribbon is packed into the filter case and moistened with oil. It traps particles of dirt that enter with the air. Over a period of time, the filter can become so loaded with dirt that it restricts the flow of air. This would prevent normal operation of the engine. Before this happens, the filter should be removed and washed in clean gasoline. It should then be reoiled and reinstalled on the carburetor.

Some carburetors use an oil-bath air filter. Different types of air filters are shown in Fig. 8.3. In the oil-bath air filter, there is a reservoir containing oil past which the incoming air must flow. As it does this, it picks up particles of oil and carries them up into the filter mesh. This tends to wash off dirt particles which drain back into the oil reservoir with the oil. On these filters, the oil must be changed at the same time that the wire mesh is washed.

8.6 Float Bowl. The purpose of the float system is to prevent the delivery of too much gasoline to the carburetor. Without the float system, all the fuel in the fuel tank would run down into the carburetor. The float system is made up of a small bowl, a float of metal or cork, and a needle valve that is operated by the float. Figure 8.4 is a simplified drawing of a float system. The sectional view of the carburetor (Fig. 8.2) shows how an actual float system looks. When gasoline from the fuel tank enters the float bowl, the float is raised. As the

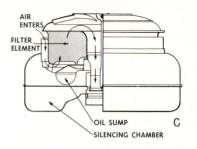

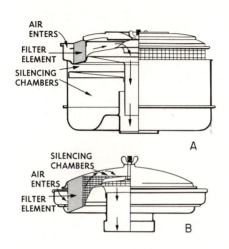

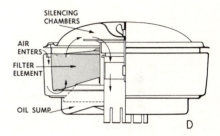

FIGURE 8.3 Air cleaners. *A* and *B*, dry type, *C* and *D*, oil-bath type.

float moves upward, it lifts the needle valve into the inlet hole (called the *valve seat*). When the gasoline is at the proper height in the bowl, the needle valve is pressing tightly against its seat so that no more gasoline can enter. When the carburetor withdraws gasoline to operate the engine, then the gasoline level in the float bowl falls, the float and needle drop down, and more gasoline can enter. In operation, the needle valve holds a position that allows gasoline to enter at the same rate that the carburetor withdraws it. This keeps the level of gasoline in the float bowl at the same height.

8.7 The Basic Carburetor. The carburetor has three basic parts besides the float system. These are the air horn, the fuel nozzle, and a throttle valve (Fig. 8.5). The throttle valve is a round plate fastened to a shaft. When the shaft is turned, the throttle valve is tilted more or less in the air horn (Fig. 8.6). Thus, it can be tilted to allow air to flow through freely, or turned to block the passage of air. This is the

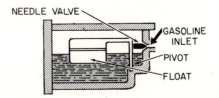

FIGURE 8.4 Simplified drawing of a carburetor float system.

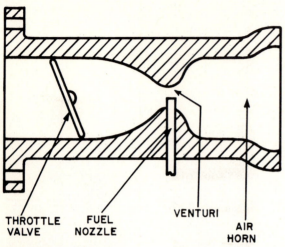

THROTTLE   FUEL   VENTURI
VALVE   NOZZLE   AIR
                                             HORN

FIGURE 8.5 Simple car-
buretor consisting of an
air horn, a fuel nozzle,
and a throttle valve.

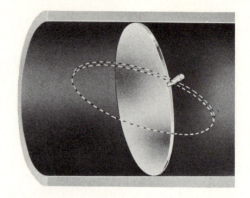

FIGURE 8.6 Throttle
valve in the air horn
of a carburetor. When the
throttle is closed, as
shown, little air can pass
through. When throttle
is opened, shown dashed,
there is little throttling
effect.

basic control because turning the throttle allows more or less air-fuel mixture to feed to the engine so the engine can produce more or less power.

The *air horn* has a restriction, or venturi, at the point where the nozzle enters it (Fig. 8.5). The purpose of the venturi is to create a partial vacuum, or low-pressure area, when air is passing through the air horn.

The following is a simplified explanation of how the venturi can create a vacuum. As air moves into the air horn, all the air molecules are moving at the same speed and are about the same distance apart. But if all the molecules are to get through the venturi, they must begin to move faster as the air enters the venturi. As the first molecule enters the venturi, it speeds up, momentarily leaving the second molecule behind. The second molecule then enters the venturi and also speeds up. But the first molecule, in effect, has a head start and cannot catch up. Therefore, in the venturi, the molecules of air are farther apart.

We know that, where molecules are relatively far apart, there is a partial vacuum, or low-pressure area. In the carburetor venturi, this vacuum is located around the end of the fuel nozzle. Then, air pressure on the fuel in the float bowl pushes fuel up the tube and out the fuel nozzle (Fig. 8.7). The fuel sprays into the passing air, mixing with it to form the air-fuel mixture that the engine needs to operate.

The more air that passes through the air horn, the higher the vacuum at the venturi and the greater the amount of fuel that feeds from the fuel nozzle. Thus, the proper proportions of air and fuel are maintained throughout the full range of throttle positions. When the throttle is partly opened, only a small amount of air flows through and only a small amount of fuel feeds from the fuel nozzle. But when the throttle is wide open, a large amount of air flows through the venturi and a large amount of fuel feeds from the fuel nozzle.

8.8 Adjusting Mixture Richness. The mixture richness can be varied by turning the adjustment knob to raise or lower the adjustment needle (Fig. 8.2). If the adjustment needle is lifted away from its seat, then the fuel passage around the needle tip is enlarged and more fuel can flow. This means that there will be more fuel and that the air-fuel mixture will be richer. But if the adjustment knob is turned to move the needle tip toward its seat, then the passage is smaller and less fuel can flow (Fig. 8.8). In this case the mixture will be leaner.

The proper richness is very important because a mixture that is too lean or too rich will not burn well and engine performance will

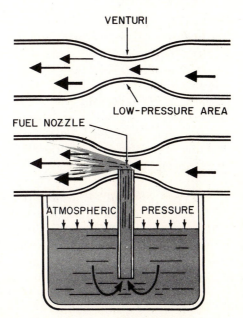

FIGURE 8.7 Top, the venturi produces a vacuum, or low-pressure area, when air flows through it. Bottom, a nozzle leading up from the float bowl will pass gasoline upward into the air stream as atmospheric pressure pushes it up toward the vacuum. (*Lawn Boy Division, Outboard Marine Corporation.*)

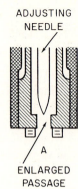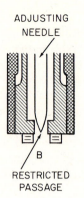

FIGURE 8.8 Carburetor adjustment needle and seat showing enlarged passage allowing more fuel to flow (*A*). Restricted passage allowing less fuel to flow (*B*).

be poor. In addition, if the mixture is too rich, not all the gasoline will burn clean. Some carbon will be left, and this carbon will soon clog the piston rings, foul the spark plug, and fill the exhaust ports. All these cause engine troubles and a weak engine.

8.9 Choke Valve.   The choke valve is located between the air filter and the carburetor venturi (Fig. 8.9). Its purpose is to help start the engine. During starting, especially when the engine is cold, only part of the gasoline will evaporate to form a combustible mixture. This means that the carburetor nozzle must deliver more gasoline to the air passing through. The choke valve has this job. The choke valve is a round plate, like the throttle valve. When the choke valve is closed, it partially blocks the air horn so that less air can get through. This produces a partial vacuum in the air horn when the engine is cranked and the piston pulls air from the air horn. This partial vacuum, added to the vacuum caused by the venturi, results in a greater vacuum at the fuel nozzle so that more fuel feeds from the fuel nozzle and the resulting mixture is enriched. After the engine starts, the choke valve must be opened to prevent delivery of an excessively rich mixture to the running engine.

8.10 Types of Small-Engine Fuel Systems.   As you can see from the above description, fuel systems for small engines are similar in many ways to the fuel systems used in automobile engines. There is a fuel tank for storing gasoline. There is a carburetor for mixing the fuel with air to produce a combustible mixture. The mixture is fed into the engine cylinder through the crankcase (on two-cycle engines) or past the intake valve (on four-cycle engines). But there are certain special features about the fuel systems you will find on small engines.

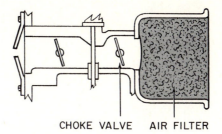

CHOKE VALVE   AIR FILTER

FIGURE 8.9 The choke valve is located between the air filter and the venturi.

We will describe these features in following sections when we cover the three basic types of small-engine fuel systems: gravity feed, suction feed, and pressure feed.

8.11 Gravity-feed Fuel System. In the gravity-feed fuel system, shown in Figs. 8.10 and 8.11, the fuel tank is located above the carburetor, and the fuel feeds down to the carburetor float bowl by gravity. Figure 8.12 shows a sectional view of a carburetor used with this type of system. The system shown in Fig. 8.1 is of this type. Fuel flows downward by gravity from the fuel tank to the float bowl when the fuel level is low in the float bowl. However, as the fuel level rises, the float rises with it and pushes the needle upward into the seat above it. When the fuel level is correct in the float bowl, the needle valve has lifted enough to close off the seat so that no more fuel will flow. As

FIGURE 8.10   Engine with a gravity-feed fuel system. (*Briggs and Stratton Corporation.*)

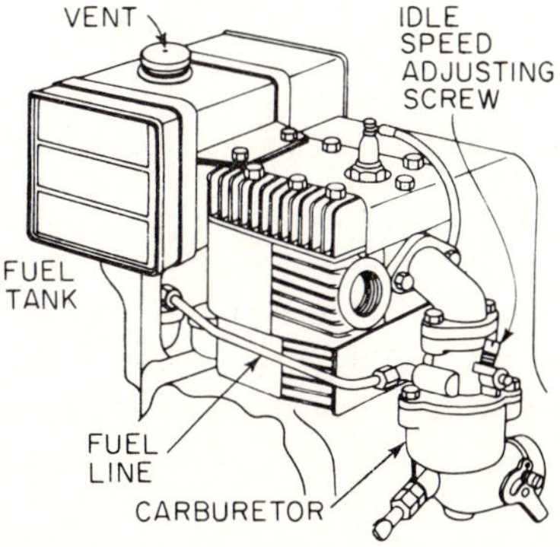

FIGURE 8.11   Gravity-feed fuel system.

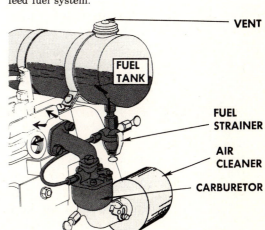

fuel is withdrawn, the float will fall to allow the needle to clear the seat and permit more fuel to flow. In operation, the float positions the needle so that the fuel flow into the float bowl just balances the fuel flow out from the float bowl. We will have more to say about the float systems of carburetors later.

The carburetor works just about the same as the carburetors used in automobiles but it is, of course, simpler in basic design and does not have an accelerator pump, a full-power circuit, and so on. The type shown in Fig. 8.10, and also in Fig. 8.13, has a choke valve in the base. When the choke valve is closed, a high vacuum develops during cranking so that more fuel feeds into the engine. That is, the mixture becomes very rich for easy starting. When the throttle is opened and the engine is running, fuel feeds past the needle valve and upward through the nozzle. It discharges through the discharge holes in the nozzle so that the ingoing air is charged with fuel. During idle, with the throttle closed, the relatively high vacuum above the throttle plate causes fuel to feed on upward through the nozzle and into the opening back of the idle needle valve. From there, it discharges into the air passing above the throttle plate.

8.12 Primer. Many small-engine carburetors have a primer instead of a choke valve. One type is shown in Fig. 8.14. The primer, when operated, supplies extra fuel to the carburetor discharge holes. The type shown in Fig. 8.14 works this way. When you press down on the

FIGURE 8.12   Sectional view of a carburetor for a small engine. (*Briggs and Stratton Corporation.*)

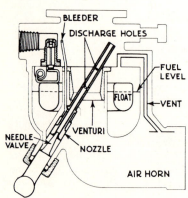

FIGURE 8.13   Sectional view of a carburetor for a small engine. (*Briggs and Stratton Corporation.*)

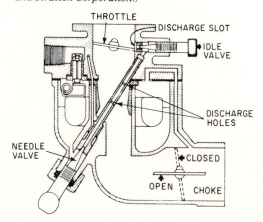

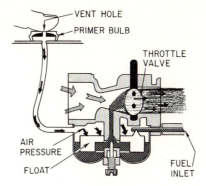

FIGURE 8.14 Bulb-type primer. When the primer bulb is pressed down, it pushes air into the carburetor float bowl, causing the float bowl to discharge fuel into the air stream through the carburetor.

primer bulb, you shut off the vent hole and force air into the float bowl. This forces fuel up through the fuel discharge hole. The primer is also called a tickler.

Another primer which is actually a small pump is shown in Fig. 8.15. This primer has a cup-shaped disk on the bottom of a spring-loaded rod. When the plunger is pushed down, fuel passes around the edge of the disk, as shown in Fig. 8.15*b*. Then, when the plunger is released, the spring pulls it up and this lifts fuel upward into the carburetor, as shown in Fig. 8.15*c*. It pours out into the carburetor air horn so that when the engine is cranked, the air passing through will be enriched for easier starting.

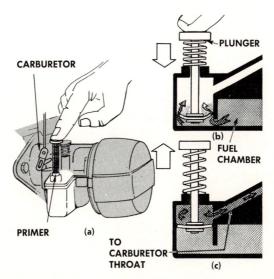

FIGURE 8.15 Primer for a small engine. When the plunger is pushed down as at (*b*), fuel passes by the cup-shaped disk. Then, when the plunger is released as at (*c*), the spring raises the disk and lifts fuel up into the carburetor.

8.13 Suction-feed Fuel System. In the suction-feed system (also called the suction-lift system), the fuel tank is below the carburetor, and fuel feeds upward directly from the fuel tank to the carburetor discharge holes. Thus, no separate float bowl is needed. Figure 8.16 illustrates how this carburetor works. Figure 8.17 is a cutaway view of an actual suction-feed carburetor. The fuel pipe from the fuel tank is connected to the two discharge holes below the point where the needle valve is located. In operation, the partial vacuum produced by the passage of air through the air horn causes atmospheric pressure to push fuel upward from the fuel tank and out through the discharge holes. Note that the throttle plate has a notch cut out of it at the top. This is for idle operation. When the throttle plate is closed, the only passage for the ingoing air is through this notch. The passing air produces a sufficient vacuum close to the idle discharge hole to cause fuel to flow so that the idle mixture contains sufficient fuel.

Some models of suction-feed carburetors have a ball check in the fuel pipe, as shown in Fig. 8.18. The purpose of this ball check is to prevent fuel in the pipe from flowing back down into the fuel when

FIGURE 8.17   Cutaway view of a suction-type carburetor. Note the slide choke, to the lower left. (*Briggs and Stratton Corporation.*)

FIGURE 8.16 Schematic view of a suction-type carburetor.

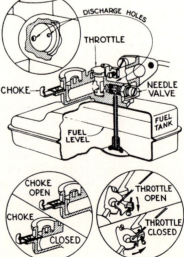

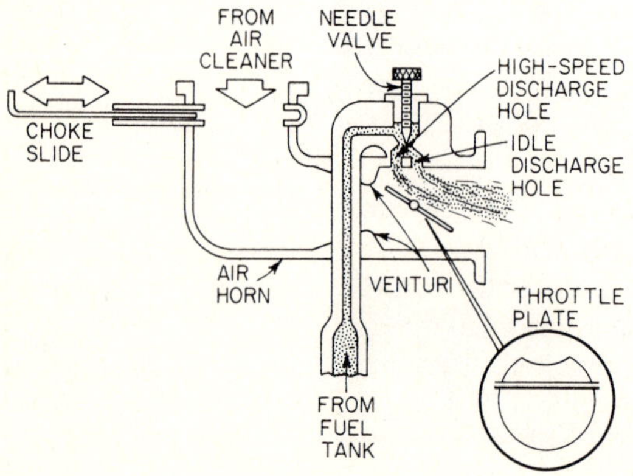

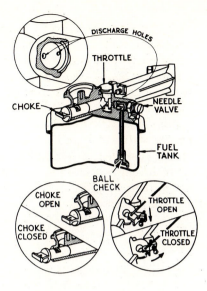

FIGURE 8.18 Cutaway view of a suction-type carburetor. Note the round slide-type carburetor choke to the lower left. (*Briggs and Stratton Corporation.*)

the engine stops. This improves subsequent starting because the fuel pipe is already filled and begins to feed fuel just as soon as the engine crankshaft is turned over.

The choke used in the carburetor, shown in Figs. 8.16 and 8.17, is typical of small-engine carburetors. It is the slide type. That is, it can be slid into the air horn to restrict air flow and thus increase the vacuum at the discharge holes. This produces additional fuel feed during cranking so that an adequately rich mixture is delivered to the engine.

Figures 8.17 and 8.18 illustrate two suction-type carburetors. The carburetor in Fig. 8.17 has a flat slide-type choke. The one in Fig. 8.18 has a round slide-type choke.

8.14 Suction-feed Carburetor with Single Control. One model of suction-feed carburetor gets along with a single control valve which provides for choking, running, and stopping the engine. A carburetor of this type is shown in Fig. 8.19 with the three positions of the valve indicated. The valve is in the shape of a half cylinder set in a hollow cylinder, as shown in Fig. 8.20. The valve can be rotated into the three positions to provide for engine control. For example, when the control valve is positioned as shown in Fig. 8.20, the air flow through the carburetor is pretty well choked off. Therefore, a high vacuum will develop on the intake stroke as the engine is cranked. This high vacuum will cause a heavy flow of fuel from the tank so the engine receives a rich mixture for starting.

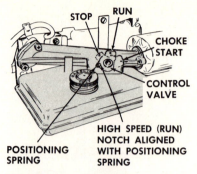

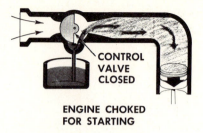

FIGURE 8.19   Exterior view of a suction-feed carburetor with a single control for starting, running, and stopping.

FIGURE 8.20   For starting, the control valve is closed so the engine is choked.

After the engine has started, the control valve is turned to the position shown in Fig. 8.21. Now, it is up out of the way, and normal engine operation results. Then, when the engine is shut off, the control valve is turned to the position shown in Fig. 8.22. In this position, the fuel pipe from the fuel tank is blocked off and no fuel can feed to the carburetor.

8.15 Suction-feed Carburetor with Diaphragm.   The suction-lift, or suction-feed, carburetor described in the previous section works satisfactorily for the smaller engines of $\frac{1}{2}$ to 3 hp (horsepower), but will

FIGURE 8.21   When the engine starts, the control valve is turned to the running position, and this leaves the control valve out of the way so air and fuel can feed into the engine.

FIGURE 8.22   To stop the engine, the control valve is turned to the closed position, as shown, so that the fuel passage from the fuel tank is closed off.

not work well for larger engines. The reason is that the vacuum will not provide sufficient fuel when the tank is nearly empty. You can see that it is easier for the vacuum to lift fuel up to the carburetor when the tank is full. But when the fuel tank is nearly empty, the fuel must be raised considerably farther, and less fuel will therefore be fed into the air passing through the carburetor. To provide for a more nearly even fuel feed, many suction-feed carburetors for larger engines have an auxiliary fuel tank or reservoir which is very similar to the float bowl in carburetors previously described.

Figure 8.23 shows a carburetor of this type with the fuel tanks removed. Note that there are two separate fuel pipes, one from the main fuel tank and one from the auxiliary tank. The arrangement is shown schematically in Fig. 8.24. Just above the main fuel tank there is a fuel pump that is operated by engine vacuum. When the piston is moving down on the intake stroke, the vacuum pulls air from the small chamber in which the pump spring is located. The pump diaphragm is pulled upward. This produces a vacuum under the diaphragm. Atmospheric pressure on the fuel in the main fuel tank then forces fuel up through the fuel pipe, as shown by the arrow. The inlet valve is opened by the vacuum, and at the same time the outlet valve

FIGURE 8.24 Schematic sectional view showing the actions taking place in the diaphragm-type, suction-feed carburetor when the piston is moving down.

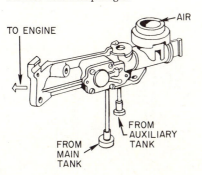

FIGURE 8.23 Exterior view of a suction-feed carburetor with a diaphragm.

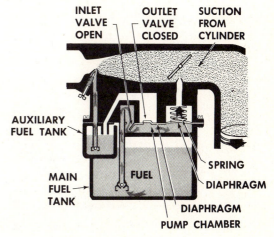

is closed by the vacuum. Fuel flows into the pump chamber. Now, when the intake stroke is completed, vacuum is lost in the carburetor. The spring can then push the pump diaphragm down. The pressure from the spring closes the inlet valve and opens the outlet valve, as shown in Fig. 8.25. As a result, the pressure can push fuel from the pump chamber upward and into the auxiliary tank. This action keeps the auxiliary tank filled so that adequate fuel can be fed in a uniform manner into the carburetor. The fuel flow is unaffected by the level of fuel in the main fuel tank.

We mentioned that the intake stroke produces the vacuum that causes the pump to work. This description fits the four-cycle engine, of course. On two-cycle engines, the vacuum would be developed when the piston is moving up on the compression stroke. The vacuum is produced in the crankcase, and the air-fuel mixture from the carburetor feeds into the crankcase past the reed valve. This vacuum operates the diaphragm of the pump. The principle of operation is the same, however, as for the four-cycle engine.

8.16 Diaphragm Carburetor. This carburetor is required for engines that are operated at various angles. The carburetors discussed previously would not work with a power saw, for example, which is held at different angles when it is used. These other carburetors depend for their operation on a float bowl in the carburetor, or a fuel tank under the carburetor. The float bowl type of carburetor has a means of keeping the float bowl filled with fuel to the proper level at all times. The tank-type of carburetor which uses suction to lift the fuel to the carburetor must have the fuel tank under the carburetor. If either of these engines is held at an angle, the fuel would run out and the engine would either be starved for fuel, or fed so much fuel it would flood and die.

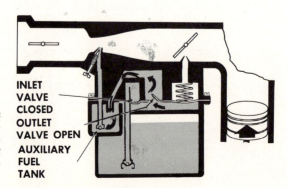

FIGURE 8.25 Schematic sectional view of a diaphragm-type, suction-feed carburetor, showing the actions when the piston is moving up.

INLET VALVE CLOSED OUTLET VALVE OPEN AUXILIARY FUEL TANK

Obviously, some other type of carburetor must be used with engines that operate at various angles and in different positions. The diaphragm carburetor is the answer here because it will provide uniform fuel feed to the engine regardless of the working position of the engine. A diaphragm carburetor is shown in simplified view in Fig. 8.26. Its operation is simple. When the piston moves down on a four-cycle engine during the intake stroke—or it moves up on the two-cycle engine—a partial vacuum is produced in the carburetor air horn. This causes fuel to discharge from the fuel reservoir into the carburetor air horn. The partial vacuum also causes the diaphragm to move up against spring tension. Then, when the vacuum is lost—piston stroke ends—the spring pushes the diaphragm down, and this creates a partial vacuum in the fuel reservoir. Atmospheric pressure then pushes fuel from the fuel tank into the reservoir to replenish the fuel withdrawn during the piston intake stroke. The action continues as long as the engine operates; providing the fuel needed to keep the engine going. Figure 8.27 shows a diaphragm carburetor.

8.17 Pressure-feed System and Fuel Pump. On engines where the fuel tank must be mounted on a level with, or below, the carburetor, a gravity-feed system will not work. A suction-feed system is often not satisfactory because it works only for small engines, as we have already mentioned. A fuel pump will deliver fuel to the float bowl of

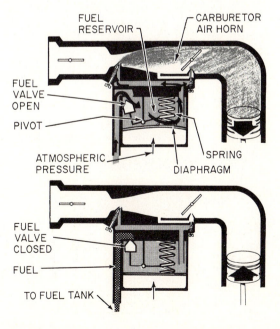

FIGURE 8.26 Schematic view of a diaphragm carburetor. Top, actions when piston is moving down on intake stroke (four-cycle engine). Bottom, actions when piston is moving up.

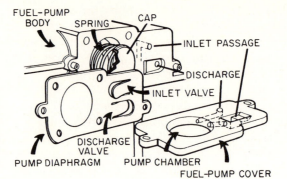

FIGURE 8.27 Partial disassembled view of a diaphragm carburetor showing the details of the pump diaphragm. (*Briggs and Stratton Corporation.*)

the carburetor regardless of their relative positions. Fuel pumps are used on all automobiles, as you know. Their purpose is to withdraw fuel from the fuel tank and pump it into the carburetor float bowl, keeping the float bowl filled to the proper level at all times.

The system using a fuel pump is called a pressure-feed system. A system of this type is shown in Fig. 8.28. A cam, or an eccentric, on the engine crankshaft forces a pump lever to move up and down, and this action produces the pumping action in the pump. In larger engines, such as those used in automobiles, the pump lever is actuated by an eccentric on the engine camshaft, rather than the crankshaft as in small engines. The principle of operation is the same, however.

Figures 8.29 and 8.30 show schematically how the fuel pump works. When the pump lever is pushed down by the lobe on the cam, it lifts the diaphragm against the pressure of the diaphragm spring, as shown in Fig. 8.29. This produces a vacuum in the pump chamber which lifts both the inlet and the outlet valves. The upward movement of the outlet valve closes it. The upward movement of the inlet valve opens it so that the vacuum will allow fuel from the fuel tank to flow into the pump chamber, as shown by the arrow.

FIGURE 8.28 Pressure-feed fuel system for a small engine. The engine has been partly cut away so the position of the pump lever on the groove in the crankshaft eccentric can be seen.

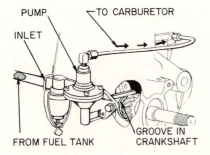

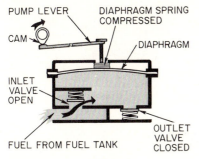

FIGURE 8.29 Action in the pump when the pump lever is pushed down by the cam lobe on the eccentric.

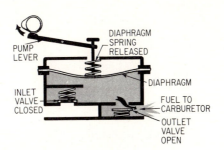

FIGURE 8.30 Action in the pump when the lobe has moved out from under the pump lever.

When the lobe of the cam moves out from under the pump lever, the diaphragm spring pushes down on the diaphragm, thus creating pressure in the pump chamber (Fig. 8.30). The pressure pushes down on both valves, causing the inlet valve to close and the outlet valve to open. The pressure then pushes fuel from the pump chamber into the carburetor float bowl. The action is continuous as long as the engine runs. When the float bowl in the carburetor becomes sufficiently filled, the float rises and lifts the needle valve up into the seat, thus shutting off any further delivery of fuel. We have already described how this works and showed a simplified float system in Fig. 8.4. When the float system refuses to take any further fuel, the diaphragm remains in its upper position, shown in Fig. 8.29, even though the pump lever releases it and the spring pressure is trying to push it down.

In actual operation, the float needle takes a position that allows just enough fuel to enter to replace the fuel leaving. The pump operates to deliver just this amount of fuel and no more. Figures 8.31 and 8.32 show the two positions of a fuel pump used in an automobile engine. This is a more accurate representation of how the inside of the fuel pump looks than is shown in Figs. 8.29 and 8.30, which are only schematic.

8.18 Governors. Where the load on the engine varies but a steady speed is required, as for instance on a lawn mower, an engine governor is needed to prevent the engine from bogging down when the going gets tough.

Basically, what the governor does is control the opening of the carburetor throttle valve. When the load is light, the engine starts to

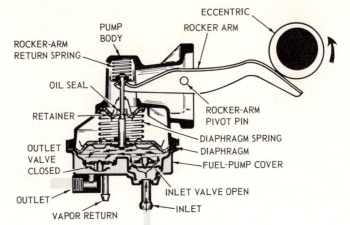

FIGURE 8.31 When the eccentric rotates so as to push the rocker arm down, the arm pulls the diaphragm up. The inlet valve opens to admit fuel into the space under the diaphragm.

speed up. As this happens, the governor causes the throttle valve to move toward the closed position. This counterbalances the speed-up tendency so that engine speed remains constant. Likewise, if the going gets heavy, as for example when the mower meets some high weeds or tough grass, then the engine tends to slow down. When this happens, the governor causes the throttle valve to open so that more air-fuel mixture can get to the engine. The engine will then develop more power to handle the heavier load without slowing down.

8.19 Types of Governor.   There are two general types of governor, the air-vane type and the centrifugal type. The air-vane type works

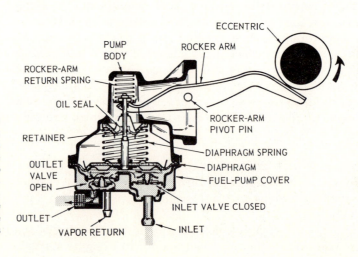

FIGURE 8.32 When the eccentric rotates so as to allow the rocker arm to move up under it, the diaphragm is released so it can move down, producing pressure under it. This pressure closes the inlet valve and opens the outlet valve so fuel flows to the carburetor.

on the flow of air from the blades on the flywheel. The centrifugal type works off a centrifugal device which is actuated by engine speed.

8.20 Air-vane Governor. The air vane is located under the flywheel shroud close to the flywheel, as shown in Fig. 8.33. It is thus in the path of the air coming off the flywheel blades. The air vane is connected by linkage to the throttle valve, as shown. When the air vane is moved, the throttle valve will open or close. There is also a spring in the linkage, as shown, which tends to pull the throttle into the opened position. When the engine is stopped, the throttle valve is open.

When the engine is running, a flow of air from the flywheel blows against the air vane, pushing it toward the right (in Fig. 8.33). As the air vane moves, the linkage tends to close the throttle. The faster the engine rotates, the stronger the air blast from the flywheel and the farther the air vane moves. Thus, the engine will not overspeed because the air-vane governor will close the throttle sufficiently to prevent it.

To take a typical example, suppose you are using a power lawn mower. You start the engine and set the throttle to run at 3,000 rpm (revolutions per minute). You wait for a few moments to allow the engine to warm up. The engine is running without load. It tends to speed up. As it speeds up beyond the preset 3,000 rpm, the air vane moves enough to partly close the throttle, thus preventing overspeeding. Now, with the engine warmed up, you start across the lawn. You hit a heavy patch of grass and this puts an extra load on the engine. It starts to slow down. Now, the air vane has less air blowing against it, and it is pulled back by the spring tension. This movement allows the throttle to open further, thus feeding more air-fuel mixture to the engine. The engine responds by putting out more power to handle the heavier load.

FIGURE 8.33 Details of an air-vane type governor. (*a*) When the engine is not running, the spring holds the throttle open. (*b*) When the engine is running, air from the blades on the rotating flywheel causes the vane to move, thereby partly closing the throttle.

8.21 Centrifugal Governor.   The centrifugal governor has a lever that is linked to the throttle through an arm and a rod. (See Fig. 8.34.) Figure 8.35 will give you a better idea of how the governor works. As engine speed increases, the two flyweights move out and push against the spool. This motion is carried to the arm so that the rod pulls down on the throttle lever and tends to close the throttle. To begin then, when the engine is started, the operator opens the throttle to get the engine speed he desires. Note that this puts a certain tension on the control spring and that the throttle is opened through the pull of the spring on the governor arm. This gives the engine the preset speed that the operator wants. Now, if the engine tends to speed up, the governor arm is turned, as shown to the right in Fig. 8.34. This tends to close the throttle to prevent overspeeding. On the other hand, if the engine tends to slow down due to heavy going, the flyweights move inward to allow the throttle to open wider. This allows the engine to produce more power. The engine thus maintains the speed at which the operator has set the throttle.

Instead of flyweights in the governor, some governors have flyballs (Fig. 8.36). The flyballs are located under a curved plate, as shown. As engine speed increases, the flyballs tend to move outward

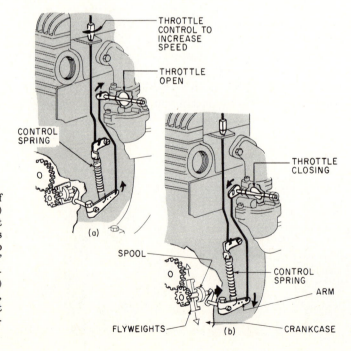

FIGURE 8.34   Details of a centrifugal governor. (*a*) When the engine is not running, the spring holds the throttle open and also holds the spool at the "in" position so that the flyweights are retracted. (*b*) When the engine runs, the flyweights move out and this causes the throttle to partly close.

and this causes them to press against the angled part of the plate, thus raising the plate. As the plate is raised, it also raises the spool. The spool, which is linked to the throttle, then causes the throttle to partly close and thus prevent overspeeding of the engine.

NOTE: **Most small engines should be operated in the high-speed range. At high speed, the engine has the capacity to adjust to a wide range of power demands. If the throttle setting is high enough, the engine is ready to start pulling hard the instant the governor calls for more power. If the throttle setting is too low, there is not enough tension on the control spring to allow the engine to start putting out full power quickly.**

CAUTION. **The governor should never be adjusted to allow the engine to run above rated speed. Even though the engine might temporarily operate at the excessive speed, and thus temporarily handle excessively heavy loads, it would quickly wear out. As a special caution, you should never operate an engine having an air-vane governor with the engine shroud removed. With the shroud off, the air flow from the flywheel is not directed against the air vane. As a result, there is no governor control. The engine could greatly overspeed and tear itself to pieces.**

A typical centrifugal governor used on a lawn mower engine is shown in Fig. 8.37. In this unit, the lower collar is fastened to the crankshaft. The upper collar is attached to the lower by a pair of pivoted links. A spring holds the two collars apart. When the engine runs,

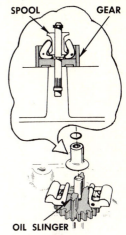

FIGURE 8.35 Details of the centrifugal governor using flyweights.

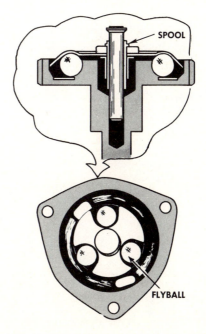

FIGURE 8.36 Details of the centrifugal governor using flyballs.

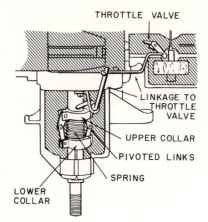

FIGURE 8.37 Sectional view of lower part of the engine, showing details of the governor, and the linkage to the throttle valve. (*Lawn Boy Division, Outboard Marine Corporation.*)

the pivoted links move out due to centrifugal force. This action moves the upper collar down toward the lower, partly compressing the spring. The faster the engine runs the greater the centrifugal force on the pivoted links and the farther down the upper collar moves by further compressing the spring. The upper collar is connected by linkage to the carburetor throttle valve. As it moves up or down, it opens or closes the throttle valve. If the engine slows down, for example, due to rough going, then the collar starts to move up. This causes the throttle valve to open and supply additional air-fuel mixture to the engine so it can produce added power. On the other hand, if the engine starts to speed up, the collar moves down to cause the throttle valve to partly close and reduce the amount of air-fuel mixture to the engine. See Fig. 8.38.

There are other types of governor. Some are mounted at the upper end of the crankshaft, above the magneto. See Fig. 8.39. All, however, work in a similar manner.

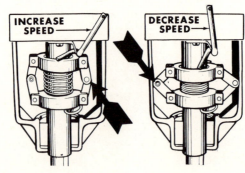

FIGURE 8.38 As the engine speed increases the pivoted links fly out, causing the control arm to move the throttle valve toward a closed position. (*Lawn Boy Division, Outboard Marine Corporation.*)

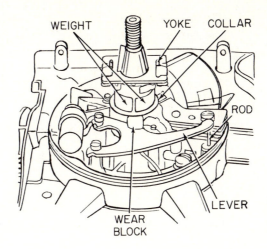

FIGURE 8.39 Governor mounted under the fly-wheel, below the magneto. (*Lawn Boy Division, Outboard Marine Corporation.*)

CHECKUP

You have learned, in studying the chapter you have just completed, that a great variety of carburetors are used on small engines. You learned the fundamentals of operation of these various carburetors, and how they work to supply the engine with the correct air-fuel mixture for good engine operation. You also learned about governors and how they work to control engine speed. The following questions will not only give you a chance to check up on how well you understand and remember these fundamentals, but also will help you to remember them better. The act of writing down the answers to the questions will fix the facts more firmly in your mind.

NOTE: **Write down your answers in your notebook. Then later you will find your notebook filled with valuable information which you can refer to quickly.**

Completing the Sentences: Test 8. The sentences below are not complete. After each sentence there are several words or phrases, only one of which will correctly complete the sentence. Write each sentence in your notebook, selecting the proper word or phrase to complete it correctly.

1. The basic elements in gasoline are (*a*) octane and oxygen; (*b*) oxygen and hydrogen; (*c*) hydrogen and carbon.
2. Increasing the compression ratio of an engine increases its (*a*) strength; (*b*) tendency to knock; (*c*) tendency to overload.
3. Two possible causes of knocking in a small engine are (*a*) late ignition timing and carbon; (*b*) high octane and compression; (*c*) early ignition timing and carbon.

4. The metal plate that can be tilted to change the air flow through the carburetor is called the (*a*) throttle valve; (*b*) air valve; (*c*) air horn.
5. The type of fuel system in which the fuel tank is located above the carburetor so it can flow down to the float bowl is called a (*a*) suction-feed system; (*b*) gravity-feed system; (*c*) pressure-feed system.
6. Two basic devices for enriching the air-fuel mixture for starting a cold engine are called the (*a*) primer and tickler; (*b*) throttle and choke; (*c*) primer and choke.
7. The type of fuel system in which the fuel tank is located below the carburetor and the fuel feeds upward directly from the fuel tank to the carburetor discharge holes is called a (*a*) suction-feed system; (*b*) gravity-feed system; (*c*) pressure-feed system.
8. The type of carburetor required for chain saws is called a (*a*) suction-feed unit; (*b*) pressure-feed carburetor; (*c*) diaphragm carburetor.
9. The two basic types of governor used on small engines are (*a*) centrifugal and pressure; (*b*) air vane and centrifugal; (*c*) air vane and pressure.
10. Is the statement "To handle heavy loads, adjust governor to increase engine power" (*a*) right; (*b*) wrong; (*c*) right part of the time.

Written Checkup

In the following, you are asked to write down, in your notebook, the answers to the questions asked or to define certain terms. Writing the answers down will help you to remember them.

1. Name the various compounds that are put into gasoline and the purpose of each.
2. Explain why knock occurs in an engine. What are two special things that could cause a small engine to knock?
3. Describe the operation of a typical gravity-feed carburetor.
4. Explain how to change the richness of the air-fuel mixture.
5. What is a primer and how does it work?
6. Explain how a suction-feed fuel system works.
7. Explain how a suction-feed system with a carburetor using a diaphragm works.
8. Explain how a diaphragm carburetor works.
9. Explain how a fuel pump works.
10. Explain how an air-vane governor works.
11. Explain how a centrifugal governor works.

Engine Starting Systems

9

In past chapters, we looked at two-cycle and four-cycle engines, found out what made them go, and how they are lubricated and cooled. In addition, previous chapters covered the ignition systems and fuel systems that engines require to make them run. But before an engine runs, it must be started. This chapter describes the different kinds of starters used in small engines. The simplest is the rope starter. You wind the rope on a pulley and give it a pull to get the engine started. The most complex is the electric starter, using either a battery or the 120 volts from the house electric wiring system. Let us look at all of these.

9.1 Types of Starters. There are four general types of small-engine starters: rope-wind, rope-rewind, windup, and electric. In the rope-wind starter, the engine has a pulley attached to the crankshaft which has a slot in the pulley flange. See Fig. 9.1. To use the rope, you hook the knot in the end into the slot, as shown. Then you wind the rope around the pulley, adjust the choke, make sure the ignition is on, and give the rope a strong pull. This spins the crankshaft to start the engine. Usually, it takes more than one pull to get the engine started.

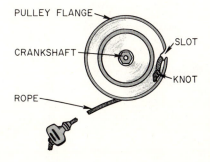

FIGURE 9.1 The rope-wind starter is the simplest of all small-engine starters. You wind the rope on the pulley and pull it to spin the engine crankshaft.

9.2 Rope-rewind Starter.   To avoid having to rewind the rope every time you attempt to start, small-engine manufacturers introduced the rope-rewind starter. This starter has the rope permanently connected and includes a recoil spring that automatically rewinds the rope on the pulley after each starting attempt. Figure 9.2 shows the basic parts of the rope-rewind starter. Figure 9.3 shows how the rope is pulled out for starting the engine.

Here is the way the rope-rewind starter works. When you pull the rope, the starter pulley is turned. This causes the pawls to fly out due to the centrifugal force on them. The pawls lock the pulley to the crankshaft so that the crankshaft is rotated when the rope is pulled out. At the same time, the recoil spring is being wound up. Note that the inside end of the spring is attached to the pulley. The outside end of the spring is attached to the starter housing. Now, after you have pulled the rope out all the way, and then released it, the spring has enough tension in it to spin the pulley back in the opposite direction. This rewinds the rope on the pulley. The rewinding takes place regardless of whether or not the engine has started. On the rewind cycle, the pawls are ineffective because they do not catch in the teeth on the inside of the crankshaft adapter. At the end of the rewind cycle, the pawls are retracted by the small springs. The starter is then ready for another starting attempt.

9.3 Windup Starter.   The windup starter is designed to reduce the amount of effort required to start a small engine. With the rope-wind

FIGURE 9.2   A partially cutaway view of a typical rope-rewind starter.

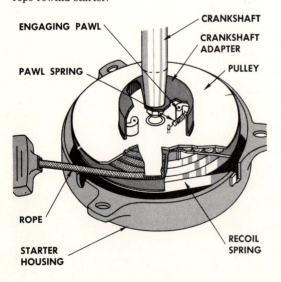

ENGAGING PAWL

CRANKSHAFT

CRANKSHAFT ADAPTER

PAWL SPRING

PULLEY

ROPE

STARTER HOUSING

RECOIL SPRING

FIGURE 9.3   How to use the rope-rewind starter. You pull it out to crank the engine. Then, you release the pull on the handle and allow the rope to rewind on the pulley.

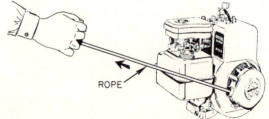

ROPE

and the rope-rewind starters, you are required to exert a strong pull to spin the crankshaft. But with the windup starter, much less effort is needed because you simply wind up a spring and then, after it is wound up, you release it. The spring then unwinds and spins the crankshaft. There are several designs of windup starters, but all operate in the same manner. Figure 9.4 shows how to use the windup starter. First, you set the release lever so it will hold the spring on windup. (Some models lock the spring when the crank is swung out.) Next, you swing the crank out and rotate it to wind up the spring. Then you return the crank to the running position. On many designs, this unlocks the release so the spring cranks the engine. On other designs, there is a separate release lever, as shown. On these, you move the release lever to release the spring and crank the engine.

Figure 9.5 shows how a typical windup starter works. It includes a crank with a ratchet and a second crank attached to one end of a spring. The other end of the spring is attached to a shaft that is part of the spring-holding mechanism. The lower end of the shaft is attached to the starter drive. The drive has a starter dog or ratchet arrangement inside the flywheel cup. When the crank handle is turned, the spring is wound up. The starter-control lever is holding the lower end of the spring so the spring is wound up. The ratchet gear and spring on the shaft of the crank prevent the spring from

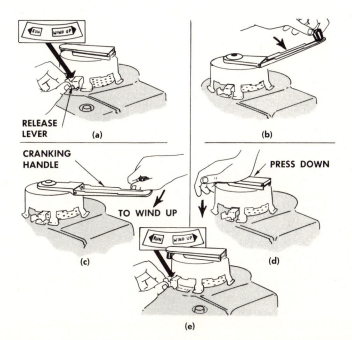

FIGURE 9.4 Here are the steps in using a windup starter. (*a*) Lock the spring by moving the control lever to "wind up." (*b*) Open the crank handle. (*c*) Wind up the recoil spring. (*d*) Fold the handle. (*e*) Release the spring by moving the control lever to "run."

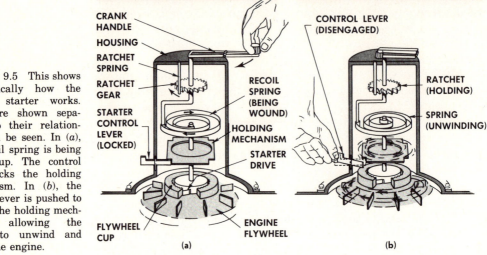

FIGURE 9.5 This shows schematically how the windup starter works. Parts are shown separated so their relationship can be seen. In (*a*), the recoil spring is being wound up. The control lever locks the holding mechanism. In (*b*), the control lever is pushed to unlock the holding mechanism, allowing the spring to unwind and crank the engine.

unwinding from the top. When the spring is wound up, and the control lever is released, the spring starts to unwind from the inside. This engages the ratchets or starter dogs inside the flywheel cup so that the flywheel is spun. This cranks the engine.

Some designs include a reduction-gear arrangement which makes it still easier to wind up the spring. See Fig. 9.6. Although it requires less effort to turn the crank, the crank has to be turned more times to wind up the spring.

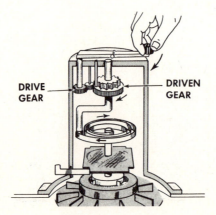

FIGURE 9.6 Some windup starters have a gear reduction which makes it easier to turn the crank although the crank must be turned more times to wind up the recoil spring.

Typical windup starters are shown in Fig. 9.7. At the top, the major parts are shown as they would appear when removed from the engine. The lower part of the illustration shows a different model of windup starter disassembled.

9.4 Electric Starters.   Electric starters for small engines are of two types. One is operated by connecting the extension cord from the starter motor to the 120-volt home wiring system, as shown in Fig. 9.8. The other type uses a lead-acid storage battery of the kind used in automobiles (Fig. 9.9).

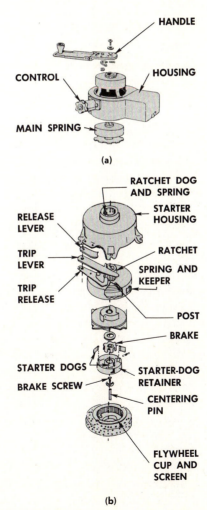

FIGURE 9.7 Assembled and disassembled views of windup starters. In (*a*), the type shown must be disassembled from the handle end. In (*b*), the type shown must be disassembled by first removing the drive mechanism.

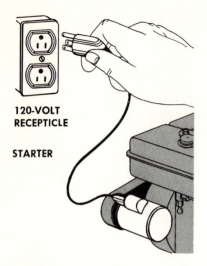

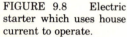

120-VOLT RECEPTICLE

STARTER

FIGURE 9.8   Electric starter which uses house current to operate.

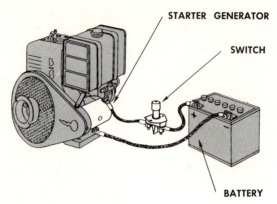

STARTER GENERATOR

SWITCH

BATTERY

FIGURE 9.9   Battery starter system. This system includes an electric starter which is also a generator to recharge the battery.

There is one caution to be observed in using the 120-volt starter. Connections should always be made at the 120-volt outlet receptacle, as shown in Fig. 9.8, and never at the starter. Reason? If you make or break the connection at the starter, the spark could ignite gasoline vapor from the carburetor and you could have a fire. Also, you should always make sure the extension cord is in good condition without frayed insulation, and that the cord is of the type having the third, or ground, lead. Remember that 120 volts is nothing to fool around with!

9.5 120-volt Electric Starter.   The 120-volt electric starter uses a small electric motor which develops sufficient power to spin the engine crankshaft and get the engine started. There are several types of electric starters. One is very similar to the type used in automobiles except, of course, it is smaller. The other two types use different sorts of drive arrangements that are special for small engines.

The purpose of the drive mechanism in the automotive type of starter is to provide a means of connecting the starter to the engine for cranking and then disconnecting it when the engine starts. In a typical automotive installation, the engine will start if it is cranked at a speed of only 100 or 200 rpm. Immediately after it starts, it will speed up to several hundred or thousand rpm. If the starter remained connected to the engine during this time, it would be subjected to very high speeds—such high speeds that it would be wrecked. So, the drive mechanism disconnects the starter from the engine before the high speed can enter the starter. In many small engines, a cone drive, a split-pulley drive, or a Bendix drive is used to connect and disconnect the starter with the engine. We will describe these in following sections.

Some small engines use a 12-volt starter-generator which not only cranks the engine for starting, but also remains connected to the engine and is driven to produce electric current for the battery and for electrical loads such as lights. We will get to this system later.

9.6 Cone-drive 120-volt Starter.   One type of 120-volt electric starter, shown in Fig. 9.10, uses a cone-shaped friction-drive clutch. To operate this starter, you press down on the switch-control button. This connects the electric motor to the 120-volt source so it spins. As soon as it gets up to speed, you then press down on the starter housing. This engages the cone-shaped drive clutch so that the flywheel and crankshaft are spun. When the engine starts, you release the starter housing and the switch-control button. The release springs lift the starter housing so the clutch disengages and at the same time the spring under the switch-control button lifts the button so that the starter motor is disconnected from the 120-volt source.

9.7 Split-pulley-drive 120-volt Starter.   This type, shown in Fig. 9.11, also engages by friction, but it does this automatically. The upper part of Fig. 9.11 shows how the split-pulley drive operates. When the starter is not operating, the two halves of the pulley are apart, as shown to the upper left. When the starter is operated, the upper half

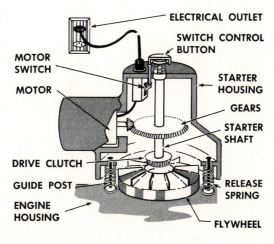

FIGURE 9.10 Details of the cone-drive mechanism used with a 120-volt starter. Parts are shown separated so their relationship can be seen.

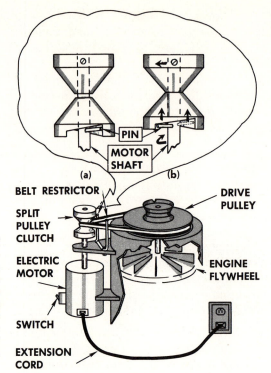

FIGURE 9.11   Details of the split-pulley-drive mechanism used with a 120-volt starter. Parts are shown schematically so their relationship can be seen. The top views, (a) and (b), show the two positions of the lower pulley half. At left, not cranking. At right, cranking.

of the pulley comes up to speed along with the motor armature and shaft because the upper half of the pulley is attached to the shaft. The lower half of the pulley, being somewhat free, does not pick up speed instantly. Instead, it lags behind because of inertia. Inertia is that property that all things have which resists change of position. Therefore, the lower pulley half is momentarily stationary. The pin in the motor shaft then pushes against the ramp in the lower pulley half, forcing the pulley half to move upward, as shown to the upper right in Fig. 9.11. Now, the sides of the two pulley halves clamp the drive belt so that the drive belt is forced to move. This causes the drive pulley located above the engine flywheel to spin, thereby cranking the engine.

After the engine starts and the motor is disconnected from the 120-volt source, the belt tension applied to the pulley forces the lower half to continue to move for a moment, and this allows the lower half to drop down to the disengaged position, as shown to the upper left in Fig. 9.11.

9.8 Bendix-drive 120-volt Starter. This starter is very much like the electric starting motors found in automobiles. At one time, the Bendix drive was used on many automotive starting motors, but today most use a form of overrunning clutch. The Bendix drive is used in some small-engine starters, however, because of its simplicity. Figure 9.12 shows a Bendix drive disassembled. Figure 9.13 shows a similar unit partly cut away. Figure 9.14 shows how the Bendix drive works.

The drive head is keyed to the motor shaft. One end of the heavy drive spring is fastened to the drive head. The other end of the spring is attached to the sleeve of the pinion-and-shaft assembly. The sleeve has threads on it, and there are matching threads on the inside of the pinion. The pinion can turn on the sleeve, and the threads cause the pinion to move back and forth on the sleeve.

When the motor is energized for cranking the engine, the armature shaft and sleeve pick up speed very quickly. The pinion lags behind, however, because of its inertia. Therefore, the sleeve turns within the pinion, and the threads force the pinion to move endwise and into mesh with the teeth on the flywheel, as shown in Fig. 9.14. As the pinion reaches the end of the sleeve, it strikes the pinion stop. Now, it must rotate with the shaft and sleeve so that the engine is cranked. The heavy spring takes up the shock of meshing.

After the engine starts, it spins the pinion faster than the armature shaft and sleeve are turning. As a result, the pinion is spun back

FIGURE 9.13 Partly cut-away view of the pinion subassembly for a Bendix drive. Note the internal threads on the pinion and the external threads on the sleeve. (*Delco-Remy Division, General Motors Corporation.*)

FIGURE 9.12 Disassembled view of a Bendix drive. (*Delco-Remy Division, General Motors Corporation.*)

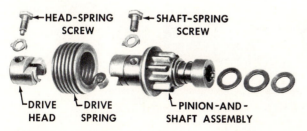

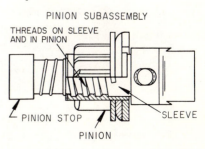

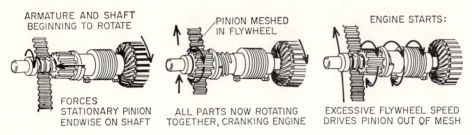

ARMATURE AND SHAFT
BEGINNING TO ROTATE

FORCES
STATIONARY PINION
ENDWISE ON SHAFT

PINION MESHED
IN FLYWHEEL

ALL PARTS NOW ROTATING
TOGETHER, CRANKING ENGINE

ENGINE STARTS:

EXCESSIVE FLYWHEEL SPEED
DRIVES PINION OUT OF MESH

FIGURE 9.14 Operation of a Bendix drive. (*Delco-Remy Division, General Motors Corporation.*)

out of mesh from the flywheel teeth. This protects the armature from excessive speed.

9.9 Battery-operated Starter.   Battery-operated starters use current from the battery to operate the starter and crank the engine. As a rule, the engine using a battery-operated starter also has a generator or alternator for restoring to the battery the current used in cranking. The generator or alternator requires auxiliary equipment to provide control, and we will discuss this later.

Figure 9.15 is a disassembled view of a typical battery-type starter for a small engine. This starter uses a modified type of Bendix drive which works in the same way as the Bendix drive previously described. Instead of a spring, this drive uses a heavy rubber cushion which takes up the shock of meshing.

FIGURE 9.15 Disassembled view of an electric starter for a small engine. (*Delco-Remy Division, General Motors Corporation.*)

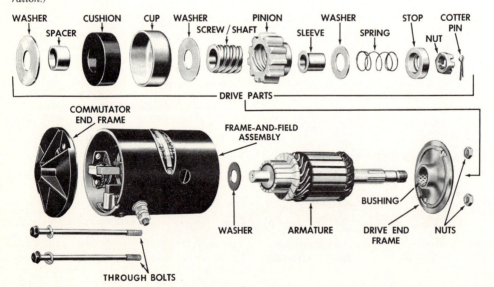

WASHER   SPACER   CUSHION   CUP   WASHER   SCREW/SHAFT   PINION   SLEEVE   WASHER   SPRING   STOP   NUT   COTTER PIN

DRIVE PARTS

COMMUTATOR END FRAME

FRAME-AND-FIELD ASSEMBLY

WASHER   ARMATURE   BUSHING   DRIVE END FRAME   NUTS

THROUGH BOLTS

The two major electrical components of the starters are the armature and the field-frame assembly, shown in Fig. 9.16. The theory of starter operation can be demonstrated with the setup shown in Fig. 9.17. If you make a pair of supports and a swinging loop out of bare copper wire, and use a horseshoe magnet and battery, as shown, you will see why the starter can develop a strong cranking effort. When you connect the two leads from the battery to the two supports, as shown, current will flow from the battery through the swinging loop. The swinging loop will swing to a horizontal position. The reason is that the current flowing through the loop produces a magnetic field around the wire. We discussed this action in Chap. 7, in our explanation of magnetism and electromagnetism. This electromagnetic field opposes the field from the permanent magnet with the result that the wire is pushed away. Therefore, the loop swings as far away from the permanent magnetic field as possible.

In a sense, this is about what happens in the starter. There are numerous heavy copper wires, or conductors, in the armature. There are two or four field coils (also called field windings) in the field-frame assembly. In the starter, the field coils and armature conductors are connected in series, as shown in Fig. 9.18. The current flowing through the field coils produces one magnetic-field pattern. The current flowing through the armature conductors produces another magnetic-field pattern. These two magnetic fields oppose each other so that the armature conductors attempt to move out from under the field-coil magnetic field. As a result, the armature spins. The circuit to the armature conductors is maintained by means of a commutator and a set of brushes. The commutator has a series of copper bars, insulated from each other but assembled in a circle. As the armature rotates, these bars come around under the brushes and make contact with them. Thus, the commutator bars connect to the brushes in succession as the armature rotates, connecting one after another of the conductors, through the brushes and the field windings, to the battery.

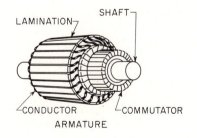

ARMATURE

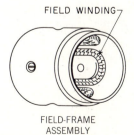

FIELD-FRAME
ASSEMBLY

FIGURE 9.16 The two major parts of an electric starter, the armature and the field-frame assembly. (*Delco-Remy Division, General Motors Corporation.*)

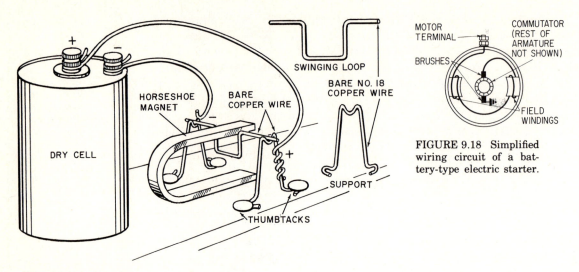

FIGURE 9.18 Simplified wiring circuit of a battery-type electric starter.

FIGURE 9.17 Simple arrangement to demonstrate motor action. When the circuit to the battery (dry cell) is completed, the loop will swing out from between the magnetic poles.

9.10 Starter-Generator. In many small-engine installations using battery-type electric starting motors, the starter may also be a generator. That is, it is a starter-generator assembly. It cranks the engine to start it, and then when the engine is running, it produces current that puts back into the battery the current taken out by the starting cycle.

Figure 9.19 shows a typical wiring circuit for a starter-generator system. It includes a generator regulator which we will describe later. The starter-generator is connected by a vee belt to the engine, as shown in Fig. 9.20, and is continuously connected during both cranking and generating. The starter-generator has two sets of field windings, one for cranking the engine and the other for producing current. When the starter switch is closed, battery current flows through the starter field windings. These windings are made of heavy copper wire so that a heavy current can flow from the battery through them. This produces a strong magnetic field which results in a strong cranking effort. The armature is spun and the engine crankshaft is rotated so that the engine starts.

Then, the operator opens the starter circuit by opening the starter switch. This opens the starter field windings so starter action is ended. Now, as the engine comes up to speed and drives the starter-generator, the generator begins to produce current. A magnetic field is produced in the generator by the generator field windings, which are made up of relatively light copper wire. These windings are shunted, or connected across, the armature and use up a small part

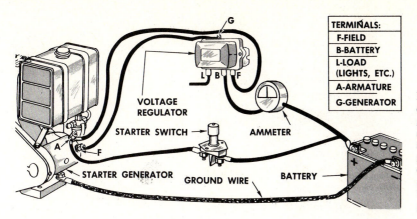

TERMINALS:
F-FIELD
B-BATTERY
L-LOAD
(LIGHTS, ETC.)
A-ARMATURE
G-GENERATOR

FIGURE 9.19 Wiring circuit of a typical starter-generator system. The starter-generator not only starts the engine, but also generates current to charge the battery. The system includes a regulator to control the generator.

of the current that the armature produces. This creates a magnetic field in which the armature spins. The armature windings which have been serving as starter windings during the starting cycle now begin to serve as current producers.

Whenever a wire conductor is moved in a magnetic field, current is produced in the wire. This is just the reverse of the starter action. When the starter works, it is due to the fact that the magnetic field produced by the field windings opposes the magnetic field produced by the armature windings. The result is that the armature spins. However, when the armature is spun in a magnetic field, current is induced in the armature windings.

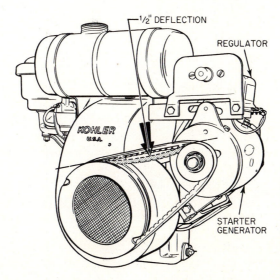

FIGURE 9.20 Belt drive for a starter-generator. Proper belt tension is indicated as allowing 1/2 in. deflection and is adjusted by moving the starter-generator toward or away from the engine. (*Kohler Company.*)

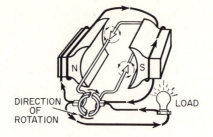

FIGURE 9.21 Simplified schematic diagram of a generator. Heavy arrows show direction of current flow, and light circular arrows indicate direction of magnetic fields around the conductors.

DIRECTION OF ROTATION

LOAD

We can sum all this up by saying that the starter uses current to produce energy to crank the engine. The generator uses energy from the running engine to produce current.

Figure 9.21 shows a generator in simplified form. It has only a single loop in the armature. The actual armature has many loops. See Fig. 9.22, which shows a disassembled generator. As the armature rotates, the loops, or conductors, are carried around in the magnetic field, and they must cut through the magnetic field. It is this cutting through the magnetic field that produces current in the conductors. The current produced is carried out from the armature through the commutator and the brushes.

The current passes through the cutout relay and regulator on its way to the battery. These units will be described in the following section.

Figure 9.23 shows schematically two variations of the basic system, one using a starter solenoid. The purpose of the solenoid is to make it possible for the starter switch to be located some distance away from the battery and starter. This reduces the length of heavy cable needed to make the circuit between the battery and starter. Only a light wire is needed between the switch and solenoid because the

FIGURE 9.22 Disassembled generator. (*Ford Motor Company.*)

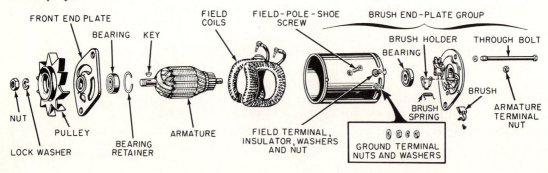

FRONT END PLATE

BEARING   KEY

FIELD COILS

FIELD-POLE-SHOE SCREW

BRUSH END-PLATE GROUP

BRUSH HOLDER   THROUGH BOLT

BEARING

BRUSH

NUT

PULLEY

BEARING RETAINER

ARMATURE

FIELD TERMINAL, INSULATOR, WASHERS AND NUT

GROUND TERMINAL NUTS AND WASHERS

BRUSH SPRING

ARMATURE TERMINAL NUT

LOCK WASHER

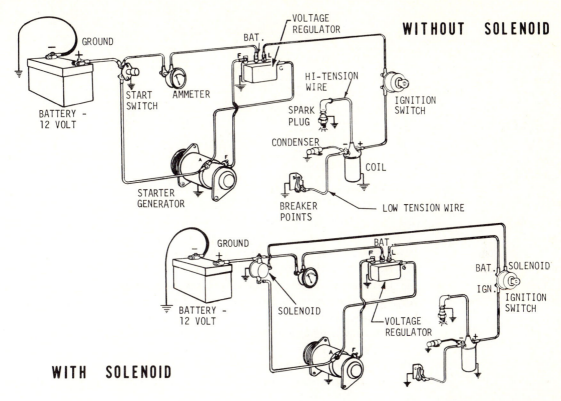

FIGURE 9.23 Wiring diagrams of two types of starter-generator systems, one without a solenoid and the other with a solenoid. (*Kohler Company.*)

solenoid needs only a small amount of current to make it work. (The bigger the current flow, the heavier the wire needed to carry it.) When the switch is turned to "Solenoid" for starting, the solenoid is connected to the battery and it produces a magnetic field. This magnetic field pulls in an iron plunger which forces heavy contacts to close. These heavy contacts make a direct connection from the battery to the starter so the engine is cranked. Note that the system shown in Fig. 9.23 includes an ignition system. This is a battery ignition system which is discussed in Chap. 7.

9.11 Current-voltage Regulator. The current, as it comes from the generator, passes through the voltage regulator. The purpose of this unit is to prevent excessive generator voltage and current, and to protect the system from battery discharge through the generator when the engine is not running. There are actually three separate devices in one in the regulator. See Fig. 9.24. These are a cutout relay, a voltage regulator unit, and a current regulator unit. Let us see how they work.

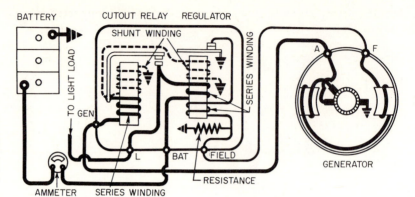

FIGURE 9.24 Wiring circuit of a combination current-voltage regulator unit and a cutout relay. (*Delco-Remy Division, General Motors Corporation.*)

1. Cutout Relay. The cutout relay has two windings, one connected or shunted across the generator, the other connected in series with the generator charging circuit. As the generator starts to rotate, its voltage builds up. This increasing voltage causes the shunt winding to produce an increasing magnetic field. When the voltage is high enough to charge the battery, the magnetism is strong enough to pull the upper contact point down and close the circuit from the generator to the battery. Current can now flow through the series winding of the cutout relay, and through the series winding of the regulator unit, and on to the battery.

When the generator slows down or stops, current starts to flow from the battery back to the generator. This reverse current, flowing through the series winding of the cutout relay, produces a magnetic field that is in the reverse direction from the magnetic field produced by the shunt winding. As a result, these two fields buck each other, and the resulting field is too weak to hold the upper contact down. The spring on this upper contact pulls it up, thus separating the points and opening the circuit between the generator and battery. This keeps the battery from discharging back through the generator.

2. Voltage Regulator Unit. The upper of the three windings on the regulator unit is a shunt winding. Voltage from the generator is impressed on this winding. As the generator increases in speed, the voltage goes up. If this voltage were allowed to continue to rise, it could under some circumstances, become so high that it would damage the system. However, the voltage regulator unit prevents this because when the voltage has gone up as far as it should, the magnetism from the shunt winding is great enough to pull the lower contact point of the regulator down. When this happens, a resistance is connected into the generator field circuit. Trace the circuit in Fig. 9.24. Note that the field circuit is connected between the left-hand brush and the terminal marked FIELD on the regulator. When the regulator points are closed, the field circuit is connected directly through these points to ground, which is the other side of the circuit. This allows full field current to flow so that full generator output can develop. However, when the contact points are separated, the generator field circuit must go to ground through the resistance. This reduces the amount of current that can flow through the field winding, and this reduces the generator output.

The voltage will fall off rapidly with the contact points open. Within a small fraction of a second, it will have dropped so much that the shunt winding of the regulator will become very weak. Its magnetism will drop so low that it can no longer hold the lower contact down. The lower contact is released so it moves up and directly grounds the generator field circuit again. Now, full generator output can be produced and the whole cycle is repeated. Actually, the contacts will open and close several hundred times a second, with the result that the resistance remains in the generator field circuit just enough of the time to prevent excessive voltage.

The series winding at the bottom of the regulator unit is connected in series with the generator field circuit. It has the job of speeding up the vibration of the contacts. When the contacts are closed, the field current flowing through this series winding adds a small amount of magnetism to help pull the upper contact down. Then, when the contacts are separated, current stops flowing through the series winding so it loses its magnetism. The total magnetism holding the lower contact down therefore drops more rapidly, hastening the closing of the contacts. The result is faster vibration of the contacts and a more even voltage.

3. Current Regulator Unit. The current regulator unit uses the generator current flowing through the heavy series winding of the regulator unit for control. When this current goes up too high so that there is danger of overloading the generator and causing it to burn out, this series winding has enough magnetism to pull the lower contact down and separate the contacts. Now, the generator field must go to ground through the resistance and generator output drops off. When this happens, the magnetism of the series winding is too low to hold the lower contact down and it moves up, directly grounding the generator field again. The result is a vibrating action that prevents excessive generator output. Helping to speed the vibration of the contacts is the small series winding through which field current flows. We have already explained how this works.

4. Combined Action. Note that the regulator provides a double control; it prevents high voltage, and it prevents high current. There are times when the battery may be fully charged so its charging voltage is high. This tends to push the generator voltage too high. Therefore, the voltage regulator unit comes into action to prevent this. At other times, when the battery is low and there is a heavy electrical load connected, such as lights, the generator tries to put out a heavy current. The current regulator prevents this by holding the maximum output of current to a safe value. This protects the generator from overworking itself and thus burning out.

The cutout relay connects the generator to the battery when the generator is ready to charge the battery, and disconnects it when the engine, and generator, slow down or stop. This prevents the battery from discharging back through the generator, which would ruin the generator.

9.12 Automotive-type Current-and-Voltage Regulator. In automobiles, the two regulator units—the current regulator and the voltage regulator—are separated into two actual units, as shown in Fig. 9.25. The operation is the same, but the separate units provide more accurate control of the system. This is important in the automotive-type systems because the generators used can produce a high output of many amperes of current. They therefore demand better control than the system shown in Fig. 9.24 could produce.

Today, this type of system is no longer used in automobiles because automotive manufacturers have switched over to a different

FIGURE 9.25 Wiring circuit of an automotive-type current and voltage regulator, with cutout relay. In this unit, the two regulator units are separate. (*Delco-Remy Division, General Motors Corporation.*)

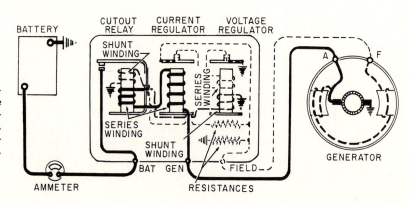

type of charging system. Now, alternators are used. Alternators are a lot like magnetos. If you remember what we said in the section on magneto ignition, you will recall that magnets moving past a stationary coil, or winding, produce a flow of current in the winding. The alternator uses the same principle.

9.13 Small-engine Alternators. In automobiles, the alternator is a sizable unit mounted on one side of the engine and driven by the fan belt. Some small engines also use alternators to furnish current for charging the battery and supplying current for external loads, such as lights. These alternators, however, are not separately mounted but are built into the engine itself. As a rule, they are combined with the flywheel magneto and use the same magnets and are called flywheel alternators.

Before we go into detail on small-engine alternators, let us review briefly how the magneto works. The flywheel has a series of magnets which are whirled past windings in the stator, or stationary part of the magneto. This produces voltage and current flow from the stator windings. The current flows through a set of contacts and the primary winding of the coil. When the contact points are separated, the current stops flowing and the magnetic field in the coil collapses. It is this collapse which produces the high voltage in the coil secondary which fires the spark plug.

The alternator operates on a similar principle and, in fact, can use the flywheel and the same stator assembly as the magneto. (But the stator assembly has different coils for the magneto and the alternator.) Figure 9.26 shows a typical installation. The engine has been partly cut away in the illustration to reveal the alternator coils and rectifiers. Figure 9.27 shows how the alternator works. The magnet

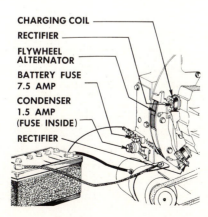

CHARGING COIL
RECTIFIER
FLYWHEEL ALTERNATOR
BATTERY FUSE 7.5 AMP
CONDENSER 1.5 AMP (FUSE INSIDE)
RECTIFIER

FIGURE 9.26 An electric starter system using a battery-type starter and an alternator mounted on the engine under the flywheel to charge the battery.

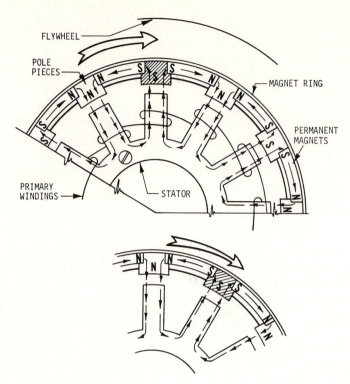

FIGURE 9.27 Schematic diagram showing how the alternator produces alternating current in the stator windings. (*Kohler Company.*)

ring mounted on the flywheel is made up of permanent magnets. When they spin past the coils on the stator, they produce a constantly changing magnetic field in the cores on which the stator windings are assembled. This means that the magnetic field is constantly moving through the windings and that current is therefore being induced in the windings.

The current induced is alternating, just like the magnetic field. That is, the current flows first in one direction and then in the other. Alternating current is of no value in charging the battery, so this must be changed to direct current, which flows in one direction only. The device which produces this action is called a rectifier. The rectifier uses a series of diodes which are electronic devices that will allow current to flow through them in one direction only. Thus, the alternating current, which flows first in one direction and then in the other, is changed to direct current, which flows in one direction only.

Let us now examine some specific flywheel alternator systems and see how they operate. The units we will look at vary in size from a small alternator that produces no more than 3 amps, to a larger unit,

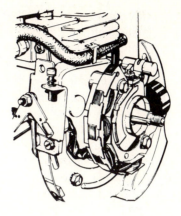

FIGURE 9.28 A 3-amp flywheel alternator. (*Tecumseh Products Company.*)

used on a two-cylinder small engine, that will produce up to 30 amps.

Figure 9.28 shows a 3-amp flywheel alternator mounted on a small engine. Figure 9.29 is the wiring diagram for this unit. Note that the system requires an ignition-starter switch which is operated by a key. The key is turned all the way for starting, and this connects the solenoid to the battery, through the switch, so that the solenoid makes a direct connection between the battery and starter. The engine is then cranked and, when it starts, the key is turned back to the engine-operating position. In this position, the circuit from the switch

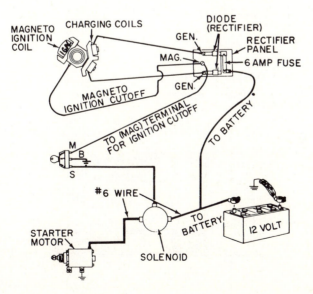

FIGURE 9.29 Wiring circuit of a 3-amp flywheel alternator which includes a flywheel magneto. (*Tecumseh Products Company.*)

to the magneto is open. This is the running position. When the engine is turned off, the key is turned to the off position, and this grounds the magneto so it stops producing the high-voltage surges needed to fire the spark plug. The engine stops.

Figure 9.30 shows the rectifier panel used with this alternator. It includes a pair of diodes to change the alternating current from the alternator to direct current. It also includes a fuse which will blow out in case the current goes dangerously high.

Figure 9.31 shows the wiring circuit for a 7-amp alternator system. This system includes lights and a double-pole switch which cuts the alternator output to about 3 amps when the lights are off. When the lights are turned on, the connections to the alternator are changed so that it can produce about 7 amps, which is enough to handle the added load and still provide a small charging current to the battery.

Figure 9.32 shows a variation of the system which includes an electronic rectifier-regulator. This unit not only converts the alternating current to direct current, but it also regulates the alternator output to suit the operating conditions. That is, it allows a high alternator output if the battery is low and needs charging, and cuts the output down as the battery comes up to charge and needs less charging current. It also allows the alternator to increase in output when the lights or other accessories are turned on.

Figure 9.33 is the wiring circuit of a 30-amp alternator used on a two-cylinder engine. Note that this system does not use a magneto but has a battery ignition system using an ignition coil and a set of breaker points somewhat like those found on automobiles. The system also includes an oil-pressure indicator light much like the indicator light used on many automobiles.

9.14 Battery. The battery we have been mentioning in past sections is a lead-acid type battery. This is the type of battery used in auto-

FIGURE 9.30 Rectifier panel for the 3-amp alternator. (*Tecumseh Products Company.*)

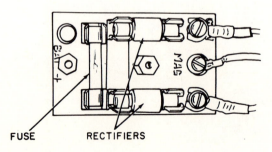

FUSE   RECTIFIERS

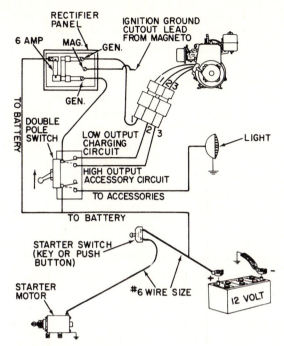

FIGURE 9.31 Wiring circuit for a 7-amp flywheel alternator. (*Tecumseh Products Company.*)

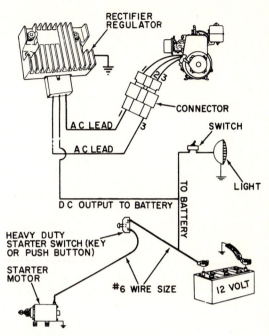

FIGURE 9.32 Wiring circuit for a 7-amp flywheel alternator which uses an electronic rectifier-regulator to control alternator output. (*Tecumseh Products Company.*)

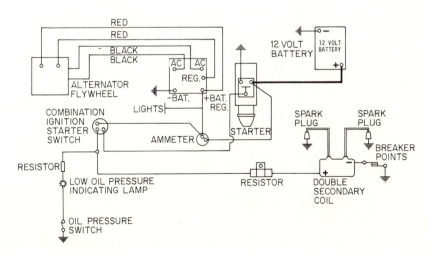

FIGURE 9.33 Wiring circuit for a 30-amp flywheel alternator using a battery ignition system. (*Kohler Company.*)

mobiles. It is a rechargeable battery. That is, when it is run down or discharged after current has been taken out of it for a while, it can be restored to a charged condition by putting current back into it in the reverse direction. Discharging current flows from the battery in one direction. Changing current flows into the battery in the opposite direction.

A typical automotive battery is shown in phantom view in Fig. 9.34. This battery has six cells, and each cell develops 2 volts. The cells are connected in series so the voltages add to produce 12 volts. Each cell contains several lead plates, kept electrically separated by separators made of wood, rubber, or glass mat.

A plate starts out as a gridwork of metal such as shown in Fig. 9.35. This is called the plate grid. To the grid is added a paste made of lead oxide. This completes the plate, ready for installation in the battery. A number of these plates are soldered, or lead-burned, to a battery post. This makes up a plate group, shown in Fig. 9.36. A second plate group is made up, using different materials in the plates, and then the two groups are put together, as shown in Fig. 9.37, to form a battery element.

A battery element goes into each cell of the battery. The cells are separated from each other by partitions in the battery case. Heavy lead connectors electrically connect the elements.

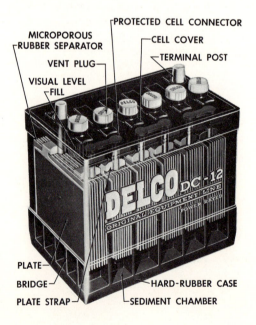

FIGURE 9.34   Phantom view of a 12-volt battery, showing internal construction. (*Delco-Remy Division, General Motors Corporation.*)

FIGURE 9.35 Battery plate grid. The circle is an enlarged section to show the shape of the grid.

FIGURE 9.36 Battery plate group.

After all cells have been completed and their covers put on, the cells are filled with sulfuric acid mixed with water. This combination of sulfuric acid and water, called the electrolyte, is essential to the operation of the battery. When the battery is charged, the electrolyte is about 40 percent sulfuric acid and about 60 percent water. As a battery produces current, or discharges, the sulfuric acid, in effect, goes into the battery plates. This action changes the lead into the plates into lead sulfate. At the same time, the electrolyte changes, in part, into water. A fully discharged battery, for example, will have

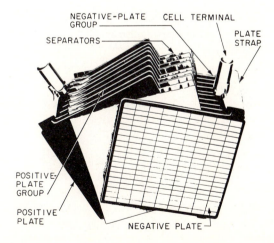

NEGATIVE-PLATE GROUP

CELL TERMINAL

SEPARATORS

PLATE STRAP

POSITIVE-PLATE GROUP

POSITIVE PLATE

NEGATIVE PLATE

FIGURE 9.37 Partly assembled battery element.

only about 10 percent sulfuric acid and about 90 percent water.

Actually, this is one way of testing the state of charge of a battery. That is, of determining how nearly discharged, or run down, it is. A device is used to find out how much sulfuric acid is still in the electrolyte. This device is called a hydrometer, and we will explain how to use it later on. There are other battery testers, and we will also describe these.

> CAUTION. **Sulfuric acid is a powerful acid and can be very dangerous if carelessly handled. If it gets on your skin, it can cause painful burns. If it gets into your eyes, it can put your eyes out! And it will eat almost anything it comes into contact with—leather, cloth, iron, steel, and so on.**

> CAUTION. **Another caution is that the battery produces explosive gases when it is being charged. These gases, hydrogen and oxygen, can be set off by a spark or an open flame and, if they explode, they can blow the top of the battery off with terrible results. So never have an open flame near a battery that is being charged, either on or off the engine.**

CHECKUP

In the chapter you have just completed, you learned about small-engine starting systems. You learned how the rope-rewind and the windup mechanical starters worked, and then studied the various kinds of electric starters used on small engines. You found out that some electric starters are also generators that put back into the battery current taken out to crank the engine. You learned about the regulators used to control generators, and also about the flywheel alternators used on some small engines. The following questions will not only give you a chance to check up on how well you understand and remember these fundamentals, but also will help you to remember them better. The act of writing down the answers to the questions will fix the facts more firmly in your mind.

NOTE: Write down your answers in your notebook. Then later you will find your notebook filled with valuable information which you can refer to quickly.

Completing the Sentences: Test 9. The sentences below are not complete. After each sentence there are several words or phrases, only one of which will correctly complete the sentence. Write each sentence in your notebook, selecting the proper word or phrase to complete it correctly.

1. The three general types of mechanical starters for small engines are the rope-wind and the (*a*) rope-rewind and electric; (*b*) 120-volt and 12-volt; (*c*) rope-rewind and windup.
2. Pulling out the rope on the rope-rewind starter winds up the (*a*) crankshaft; (*b*) recoil spring; (*c*) recoil starter.

3. On the windup starter, turning the crank (*a*) winds up the spring; (*b*) winds up the crankshaft; (*c*) unwinds the spring.
4. When connecting the 120-volt starter, always make the connections at the (*a*) starter; (*b*) engine; (*c*) 120-volt outlet.
5. The 120-volt starter uses either a split-pulley drive, a (*a*) cone drive, or a Bendix drive; (*b*) spring drive, or a Bendix drive; (*c*) Bendix drive, or a windup drive.
6. What makes the pinion mesh in the Bendix drive is the (*a*) lever; (*b*) solenoid; (*c*) pinion inertia.
7. You learned in studying the starter-generator that for starting the engine and producing current (*a*) the drive pinion is used for both; (*b*) the same armature is used for both; (*c*) the same field windings are used for both.
8. The voltage regulator controls generator voltage and output by changing (*a*) engine speed; (*b*) generator speed; (*c*) generator field current.
9. The flywheel magneto and the flywheel alternator use (*a*) the same magnets; (*b*) different magnets; (*c*) different flywheels.
10. During battery discharge, the sulfuric acid, in effect, (*a*) evaporates; (*b*) produces hydrogen and oxygen; (*c*) goes into the plates.

Written Checkup

In the following, you are asked to write down, in your notebook, the answers to the questions asked or to define certain terms. Writing the answers down will help you to remember them.

1. How does the rope-rewind starter work?
2. How does the windup starter work?
3. What is the purpose of the drive mechanism in electric starters?
4. Explain how the cone drive for small engines works. The split-pulley drive.
5. Explain how the Bendix drive works.
6. Explain how the battery-operated starter works. What makes the armature spin?
7. Explain how a starter-generator works.
8. Explain how the cutout relay works.
9. Explain how the voltage regulator works. The current regulator.
10. Explain how the flywheel alternator works.
11. Describe the construction of a lead-acid battery and explain what happens in it during discharge. During charge.
12. Explain the two important cautions to observe when working with batteries.

Trouble Diagnosis of Small Engines

10

In this and following chapters, we will look at the procedures to follow to find troubles in small engines, and what to do when various kinds of trouble are found. This chapter is devoted to trouble diagnosis, or what is commonly called troubleshooting. Later chapters explain what to do when the causes of troubles are found.

10.1 Common Small-Engine Abuses. Small engines are built to "take it." They have comparatively large crankshafts and bearings, for example, considering the horsepower they produce. A minimum requirement to meet government specifications is that these small engines should operate at full load and top speed for 1,000 hr. This may not seem like many hours at first glance, but consider this. Suppose you used your power lawn mower 4 hr a week for 6 months. This is only about 100 hr of operation a year. At this rate, the engine should last you at least 10 years. Whether it lasts this long, or longer, depends a great deal on the sort of maintenance the engine gets. Some of the abuses that shorten an engine's life include:

1. *Allowing Dirt to Get into the Engine.* This will result from inadequate servicing of the air cleaner and fuel strainer, from improperly replacing spark plugs, and not replacing oil in four-cycle engines, contamination of the fuel, and so on.

2. *Failure to Check the Crankcase Oil Level on Four-Cycle Engines Often Enough.* This can allow the oil to drop too low with a resulting lack of adequate lubrication of the engine, which results in rapid engine wear and early engine failure.

3. *Failure to Feed the Two-Cycle Engine the Proper Oil-Fuel Mixture so that the Engine is Inadequately Lubricated, Wears Rapidly, and Fails Early.*

4. Overloading the Engine so that It Works Too Hard and Wears Out Fast.

5. Running the Engine Too Fast.   Some people change the governor setting so the engine will run faster and handle heavier loads. The engine wears out rapidly under these conditions. Overspeeding the engine is a sure way to shorten engine life.

6. Failure to Properly Store the Engine during the Off Season.   Many engines are used on machines that are in use only part of the year. When they are not to be used for several weeks or months, engines should be prepared for the idle period. Failure to do this can lead to early engine failure.

Of course, you, as an engine technician, cannot see to it that all your customers give their small engines the ideal treatment. But when they bring their small engines to you for repair, you should know the various things that could cause rapid engine wear and early engine failure. That is the reason we list the six possible causes of rapid engine wear above. These are by no means the only causes of small-engine abuse. One of the more common causes of trouble is that the wrong tools are used in attempted repairs or adjustments. For proper repair, the right tools should be used.

10.2 Trouble Diagnosis of Small Engines.   When you are faced with a balky engine that will not start or will not operate properly, there are certain basic checks you can make to locate the cause of trouble. The two most common complaints people have about small engines is that they will not start and that they lack power. In addition, the engine may surge, repeatedly increasing in speed and then slowing down. Also, it may gradually lose power as it is operated, or it may fire irregularly.

10.3 Engine Will Not Start.   Failure of the engine to start could be due to lack of fuel, fuel not feeding to the carburetor, carburetor not feeding fuel to the air passing through the air horn to the engine, clogged air filter, clogged exhaust ports, defective ignition system, or internal engine damage. To check out the engine and locate the trouble, proceed as follows:

1.   Make sure there is clean gasoline in the fuel tank.
2.   Be sure the vent in the fuel-tank cap is clear. If it is clogged, the engine may start, but soon stop because the clogged vent does

not permit gasoline to flow rapidly enough from the fuel tank to the carburetor.

3. Check the engine compression by slowly pulling the engine through the compression stroke with the starter. See Fig. 10.1. Be sure the ON-OFF switch is off. If the starter is of the rope-wind or rope-rewind type, you can judge the compression by the feel. For instance, if the engine spins very easily, then there is little compression, probably due to a loose cylinder head, defective head gasket, loose spark plug, cracked head or cylinder, broken piston rings or piston, or, on four-cycle engines, a defective valve that hangs open. First check the spark plug, and if it is tight, then you should look for the other causes of trouble, disassembling the engine as necessary.

If the engine uses a windup or an electric starter, you will have to judge the compression by the way the engine acts when it is cranked. With the windup starter, if release of the spring turns the engine over unusually fast or long, you can suspect loss of compression. The same thing can be said about the electric starter, except that you should be careful not to blame the engine if the battery is run down, or if the starting motor is at fault.

When checking compression, listen for unusual squeaks, squeals, scraping or knocking sounds. Any of these could mean worn bearings, scored cylinder walls or pistons, broken rings or other parts. If you hear such noises, do not try to start the engine before carefully checking engine parts, disassembling the engine as necessary to examine them.

If the engine has normal compression, it will resist the pull of the rope—or act normally when starting is attempted with the windup or electric starter. Another sign of good compression is a sucking sound when the engine is spun fast, followed by a sort of cough as the engine stops after the spin, indicating the engine is taking in air normally.

FIGURE 10.1 Check engine compression by slowly pulling the engine through the compression stroke with the starter rope.

Try to start by choking the engine, making sure the ON-OFF switch is turned on, and then cranking the engine. If the engine shows normal compression but will not start, then the ignition system or carburetor is probably at fault. Check the ignition system first by disconnecting the high-tension lead from the spark plug. Pull back the rubber hood to expose the lead clip, or put a bolt into it to get metal contact. Hold the clip or bolt about 3/16 in. from the cylinder head and crank the engine as shown in Fig. 10.2. If a strong spark jumps to the cylinder head, the ignition system is probably okay.

If no spark occurs, then the ignition system is probably at fault and it should be checked. Causes of trouble could be dirty or worn contact points, points out of adjustment, a defective capacitor, high-tension lead, ON-OFF switch, or magneto coil.

If a spark does jump from the bolt or high-tension-lead clip, examine the spark plug to see if it can deliver the spark to the engine cylinder. Remove the plug, reattach the high-tension lead to it, and lay it or hold it against the cylinder head as shown in Fig. 10.3. Crank the engine. If no spark jumps the gap, the spark plug is probably at fault. Examine it for cracks, black sooty deposits on the porcelain or elec-

FIGURE 10.2   Check the ignition system by disconnecting the high-tension lead from the plug and put a bolt into it to get metal contact. Hold the lead with the bolt in it about 3/16 in. from the cylinder head while cranking the engine.

FIGURE 10.3   Checking the plug for spark.

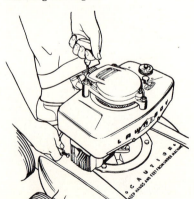

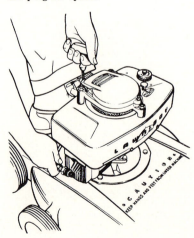

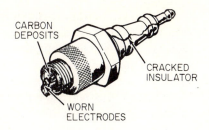

FIGURE 10.4 Defective spark plug showing cracks, carbon deposits, and worn electrodes.

trodes, burned electrodes, or wide gap. See Fig. 10.4. Any of these could prevent a good spark.

There is one other point to notice when checking a spark plug just removed from an engine that has been cranked in an attempt to start: If the end is wet with gasoline, then chances are fuel is getting to the engine. Put your finger over the spark plug hole in the head and crank the engine with the choke on. See Fig. 10.5. If your finger gets wet, it is added evidence that fuel is getting through.

If the end of the plug, or your finger does not get wet with gasoline, then the carburetor is at fault. The trouble could be due to clogged lines or nozzle, incorrect adjustment, or defective float system.

Try to adjust the carburetor. If this fails, then the carburetor will have to be removed for disassembly and repair. A typical adjustment is as follows: Turn adjusting knob, shown in Fig. 10.6, down (clockwise) to bring main nozzle needle down on its seat. Do not turn down

FIGURE 10.6 Location of adjusting knob and choke linkage.

FIGURE 10.5 Using thumb to feel if fuel is entering the cylinder head.

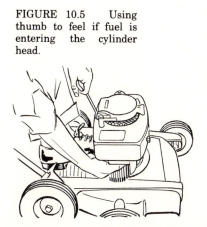

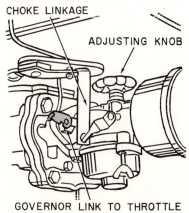

tight, since this might damage the seat or the needle and they would require replacement. Back off the knob two full turns. Close the choke and crank to see if gasoline appears in engine, using a finger on the plug hole to check. If gasoline now appears, replace plug and try to start. If the engine starts, open the choke in a normal manner as the engine runs and warms up. If the engine runs roughly, it may be getting too much gasoline. Turn needle knob in to produce a leaner mixture. After the engine is warmed up, turn the needle knob in until the engine begins to die from an excessively lean mixture. Then back out about one-fourth turn. This should be the best adjustment.

A later chapter describes in detail the adjustment procedures for various types of carburetors.

10.4 Engine Starts but Lacks Power.   A common cause of this trouble, in two-cycle engines, is clogged exhaust ports. Carbon that forms as a result of the combustion action in the cylinder often cakes up around the exhaust ports, as shown in Fig. 10.7. As this buildup continues, the engine is less and less able to exhaust burned gases and thus less fresh charge can enter the engine cylinder. This means that engine power is lost. After some time, if the accumulations are not removed, the engine will barely run. To remove the accumulations, take off the exhaust muffler. Turn the engine flywheel so the piston covers the exhaust ports. Then use a screwdriver or a hardwood scraper, as shown in Fig. 10.8, to carefully scrape away the carbon accumulations. The piston will keep particles from falling into the cylinder, where they could cause trouble. Be extremely careful to avoid scratching the piston. Blow out all loose particles from the ports. This procedure is described in more detail in a later chapter.

If clogged exhaust ports are not the cause of lack of power, then check and adjust the carburetor. Chances are the carburetor is supplying an overrich or overlean mixture.

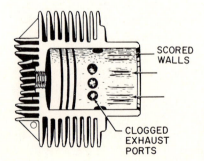

FIGURE 10.7   Cutaway view of cylinder head showing scored walls and clogged exhaust ports, caused by carbon deposits in the exhaust ports.

SCORED WALLS

CLOGGED EXHAUST PORTS

FIGURE 10.8   Use a hardwood scraper or screwdriver to remove carbon from the exhaust ports.

If the lack of power is not due to clogged exhaust ports or faulty carburetor action, then the trouble probably is in the engine itself. It could be due to worn pistons or cylinders, or worn or broken rings. One other possible cause should be considered, and this is a defective reed valve in the crankcase. If this valve is not seating properly, it may not hold compression in the crankcase, with the result that not enough air-fuel mixture will be retained in the crankcase. The charge going to the combustion chamber on intake will not be enough for the engine to develop full power. If the reed valve is warped or bent so it does not lie flat against the inlet holes, it should be replaced.

10.5 Engine Surges.   If the engine surges, that is, it repeatedly speeds up and slows down, the trouble probably is in either the carburetor or the governor. Try readjusting the carburetor as already explained. If this does not cure the trouble, then check the governor. Things to look for in the governor are binding of the linkage between the governor and the throttle valve, a weak or damaged spring, worn or binding governor parts.

If engine speed is not correct, it can be adjusted on some models by bending the linkage between the governor and the throttle valve. On other models, adjustment is made by changing governor springs. Do not attempt to change speed by stretching a spring. Chances are the spring will not hold its new set and engine operation will be unsteady. A following chapter describes governor service in detail.

10.6 Engine Loses Power.   If the engine starts off okay but gradually loses speed as it warms up, the most likely cause is in the fuel system. For example, the vent in the fuel-tank cap might be clogged, or the needle in the float bowl might be stuck. In either case, too little gasoline gets through to the carburetor, and the engine slows down because it is fuel starved.

Lack of lubrication in the engine, as for instance from failure to put oil in the gasoline (two-cycle engine), might cause loss of power as the engine warms up. Chances are this would soon cause complete engine failure from seized bearings or scored cylinder walls or pistons. See Fig. 10.9.

10.7 Irregular Firing.   If the engine fires irregularly, it could be due to a weak spark or poor carburetion. Check the spark as already described. Replace coil or capacitor, clean and adjust or replace contact points, replace wires as necessary. Check and adjust the carburetor as already described.

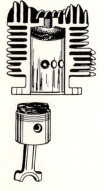

FIGURE 10.9 Scored piston and cylinder walls caused by lack of lubrication.

CHECKUP

In the chapter you have just completed, you learned the fundamentals of small-engine trouble diagnosis. This knowledge will enable you to put your finger on engine troubles without a lot of guessing. The following questions will not only give you a chance to check up on how well you understand and remember these fundamentals, but also will help you to remember them better. The act of writing down the answers to the questions will fix the facts more firmly in your mind.

NOTE: Write down your answers in your notebook. Then later you will find your notebook filled with valuable information which you can refer to quickly.

Completing the Sentences: Test 10.   The sentences below are not complete. After each sentence there are several words or phrases, only one of which will correctly complete the sentence. Write each sentence in your notebook, selecting the proper word or phrase to complete it correctly.

1. Government specifications state that a small engine should operate at full load and top speed for (*a*) 100 hr; (*b*) 1,000 hr; (*c*) 10,000 hr.
2. If the engine turns over very easily, chances are it (*a*) has carburetor trouble; (*b*) has lost compression; (*c*) needs magneto service.

3. If the engine has normal compression but does not start, check the (*a*) rings and valves; (*b*) lubrication system; (*c*) ignition and fuel systems.
4. If you do not get a spark on the spark test, chances are the (*a*) ignition system is okay; (*b*) fuel system is at fault; (*c*) ignition system is at fault.
5. A common cause of loss of power in two-cycle engines is (*a*) clogged exhaust ports; (*b*) valves hanging up; (*c*) worn bearings.
6. Loss of compression in the crankcase of two-cycle engines is probably due to (*a*) the intake valve hanging up; (*b*) the exhaust valve hanging open; (*c*) a defective reed valve.
7. If the engine starts okay but gradually loses power as it warms up, chances are the trouble is in the (*a*) ignition system; (*b*) fuel system; (*c*) governor.
8. Irregular firing is probably due to trouble in the (*a*) ignition or fuel system; (*b*) lubrication system; (*c*) governor.

Written Checkup

In the following, you are asked to write down, in your notebook, the answers to the questions asked or to define certain terms. Writing the answers down will help you to remember them.

1. Make a list of the common small-engine abuses that shorten engine life.
2. List the procedure to follow if the engine will not start.
3. Explain how to check compression.
4. Explain the procedure to follow if the engine starts but lacks power.
5. What are possible causes of the trouble if the engine surges?
6. If the engine starts okay but loses power as it warms up, what are the possible causes?
7. If the engine fires irregularly, what are the possible causes?

Maintenance and Operation of Small Engines

11

In this chapter, we discuss the maintenance of small engines, and explain how proper maintenance will prolong the life of engines. Many small engines meet an early death because they are not properly taken care of. With proper care and maintenance, small engines will give the life that was built into them. As we mentioned previously, a minimum requirement to meet government specifications is that small engines should operate at full load and top speed for 1,000 hr. This is a lot of hours and means years of service for most small engines that are used only intermittently. Section 10.1 lists and describes the common small-engine abuses that shorten engine life. In this chapter, we will look at the maintenance procedures that will help to ensure good engine performance and long engine life. First, we list a number of checks that should be made periodically, and then we go into detail on how to perform the various recommended maintenance checks.

11.1 Maintenance Hints.   Let us list maintenance hints for small engines. In later sections, we will explain fully how to perform these maintenance steps, such as cleaning the air filter, cleaning the engine, and so on.

1. Cleaning Air Filter.   Clean and reoil the air filter regularly. You will usually find instructions on the engine on when and how to do this job. We will describe the procedure for various types of air filter in a later section. Usually, the recommendation calls for doing this job every 10 hr of engine operation. If operating conditions are especially dusty, clean the filter every 5 or 6 hr.

2. Nuts and Bolts.   Check the tightness of all bolts and nuts on the engine and complete machine periodically. They sometimes loosen up in service and if not retightened, parts may become damaged or lost.

3. Lubrication.   Lubricate all bearings outside the engine, as for instance the wheel bearings on a power mower. Make sure oil reservoirs are filled (as for instance, that on the chain saw lubricator).

4. Blades and Saw Teeth.   Make sure the blades (or saw teeth) are sharp and that the rest of the assembly is in good condition.

5. Fuel Tank.   It is desirable to keep the fuel tank filled. If it is allowed to sit around only partly filled, air will enter and leave the tank as the temperature changes. This will introduce moisture into the tank. It will condense and will ultimately cause severe rusting of the tank (metal tank). Not only will this damage the tank, but rust particles may get into the carburetor and cause clogging of the fuel passages.

6. Keep the Machine Clean.   Wipe it off periodically to remove oil, grass clippings, dust, and so on. Remember that collecting of such trash around the engine will act as a blanket so the engine may overheat. On mowers, clean off the accumulations of grass clippings from the inside of the housing. Section 11.3 describes in detail how to clean engines.

> CAUTION.   **Be sure the engine switch is turned off and the spark plug high-tension lead is disconnected so the engine will not start when you work on the business end of any machine. For an example of what could happen, if a rotary mower should start while you are working on the underside, you could be seriously injured.**

11.2 Maintenance Procedures.   Many of the maintenance procedures we describe here should be performed periodically by the owner or operator of the engine. Unfortunately, as we have already noted, the engine does not always get the maintenance it should with the result that its life is shortened. We will cover all the recommended maintenance procedures in following sections so that you will be

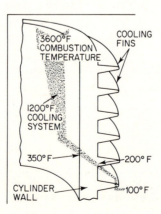

FIGURE 11.1   Heat-travel path from combustion gases to the lower cooling fins. Excess heat is lost through the cylinder wall fins.

familiar with them. One of the most important services is to clean the engine and its components such as the shroud, carburetor air cleaner, fuel strainer, and crankcase breather. It is dirt, more than anything else, that is the enemy of long engine life.

11.3 Cleaning the Engine. Most small engines are air cooled and have cylinders with a series of cooling fins. These fins provide large surface areas from which heat can be radiated. The heat flows from the inside of the cylinder, through the cylinder metal, to the fins which radiate it to the outside air, as shown in Fig. 11.1. If these fins become dirty or covered with oil, grass clippings, and so on, the heat cannot get through. The accumulations act as a blanket to hold heat in the engine. As a result, the engine becomes overheated. The oil film on the engine parts becomes less effective, or actually fails, with the result that engine parts wear rapidly and engine life is shortened. It is essential, therefore, for long engine life, that the engine be cleaned before each use, if required.

Another purpose of periodically cleaning the engine is to check for loose nuts or bolts, and loose, broken, cracked, or otherwise damaged parts. A simple way to clean the engine is to use a stiff brush and water. This will not only get into all the crevices where dirt can accumulate, but it will clean away most of the grass clippings and other trash that can cause trouble. For a complete cleaning job, you should use a degreasing compound, as explained later.

CAUTION. **Do not clean a hot engine. Wait until it is cool. If you throw water on a hot engine, you can crack the cylinder. Also, many degreasing compounds are flammable and could burst into flames when sprayed on a hot engine.**

Parts to be cleaned include the shroud—on engines so equipped—muffler and exhaust ports—on two-cycle engines—air cleaner, fuel strainer, and crankcase breather.

11.4 Cleaning the Shroud. Many small engines have fans and shrouds to direct the flow of air around the engine cylinder, as shown in Fig. 11.2. The shroud will have to be removed before the engine can be cleaned. Figure 11.3 shows one shroud arrangement. Shrouds are held in place by screws which can be taken out to allow the shroud parts to be lifted off.

NOTE: **On a few engines, it will be necessary to remove certain other parts before the shroud can be removed. These parts might include the air cleaner, muffler, spark plug wire, governor spring, or some other such minor part.**

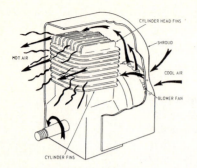

FIGURE 11.2 Action in a cooling system of a small engine which is shrouded. Arrows show flow of cooling air around cylinder and cylinder head.

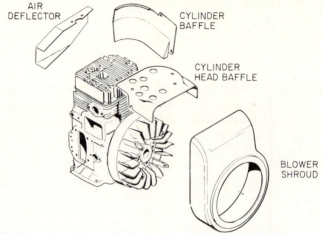

FIGURE 11.3 Small engine with the shrouds removed.

CAUTION. **Never operate the engine with the shroud and baffles removed! The shroud is there to flow cooling air over the engine. When it is off, the engine will overheat if operated. In addition, engines which have governors that operate on air flow will not function properly with the shroud off, with the result that the engine can overspeed and probably ruin itself.**

If the shroud is bent or damaged, it should be straightened, repaired, or replaced. A defective shroud can cause engine overheating, and also it might interfere with the fan or other moving part.

If the shroud is dirty and has accumulations of grass clippings or other trash, scrape it off with a putty knife or similar instrument. Use a stiff-bristled brush and solvent if necessary. Clean the air-intake screen with a brush and solvent if necessary to get rid of all accumulations of trash that could prevent normal air flow through it.

11.5 Cleaning the Cylinder and Cylinder Head. The fins on the cylinder and cylinder head should be clean to permit maximum heat transfer from the engine to the surrounding air. Three substances for cleaning the cylinder and head can be used, a degreaser, a solvent, or live steam. As a first step, use a wooden stick to scrape away all the accumulated trash and gunk. Do not use a metal tool because this will scratch the cylinder and head and encourage accumulations of dirt.

Then, use the material you have on hand to finish the cleaning job. Degreasing compound comes in pressure-spray cans or in larger containers. Solvents can be mineral spirits, kerosene, or diesel fuel. To use live steam, you need a steam generator.

While cleaning the cylinder and head, check for oil leaks which usually show up as a heavy accumulation of dirt. Check also for cracks or other damage.

Apply the solvent on the areas to be cleaned. The degreaser in the pressure can is the easiest to use. Other types can be applied with a paint brush. After about 5 min, flush off the solution with a stream of water from a hose or, if you have used an oily solvent such as kerosene, use a solution of soapy water brushed on and then flushed off.

> CAUTION. **Do not clean a hot engine. Allow it to cool first. Cold water or other liquid on the hot engine can cause the head or cylinder to crack. Some cleaning solutions are flammable and could burst into flames if sprayed on a hot engine. Also, make sure that there is adequate ventilation. Some fumes from cleaning solutions are unhealthy to breathe.**

11.6 Cleaning Muffler and Exhaust Ports. On two-cycle engines, the exhaust ports tend to become clogged with carbon accumulations, as shown in Fig. 10.7. These accumulations reduce the ability of the engine to exhaust and can sharply reduce engine power. They can also cause engine overheating which will result in other engine damage. To check for carbon accumulations, remove the muffler, as shown in Fig. 11.4. Mufflers come in a variety of sizes and shapes, but practically all are held in place by cap screws and a gasket. Removing the screws permits you to inspect the exhaust ports.

If the ports are clogged, rotate the crankshaft until the piston moves down enough to cover the exhaust ports. This protects the engine from dirt and carbon which could otherwise fall into the engine and cause damage. Use a wood stick to scrape off the carbon, as shown in Fig. 10.8. Do not use a metal tool that could scratch the piston or damage the edges of the exhaust ports. Hold the engine so

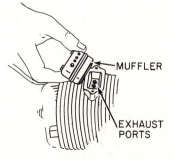

FIGURE 11.4 Removal of muffler to inspect exhaust ports of a two-cycle engine.

that the exhaust ports are pointing down while scraping. This allows the loosened carbon to fall out and reduce the chances of any of it getting into the engine. To finish the job, blow out the ports with compressed air or use a brush to make sure you have removed all the carbon. Clean the muffler in solvent. When replacing the muffler, use a new gasket if the old one appears damaged, and tighten the attaching screws securely.

11.7 Cleaning the Carburetor Air Cleaner. Carburetor air cleaners for small engines can be classified as oil-bath, oiled-filter and dry-filter. These are shown in Figs. 11.5 to 11.7. In the oil-bath cleaner, shown in Fig. 11.5, there is an oil cup in the bottom of the cleaner, as shown. Incoming air passes over the oil in the oil cup and it picks up

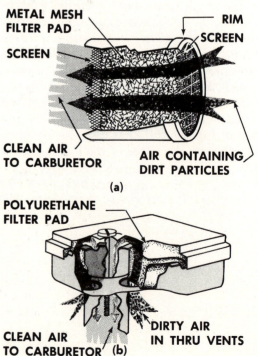

FIGURE 11.6 Oiled-filter air cleaners. (*a*) A pad of metal foil ribbons form a mesh through which the air must pass. (*b*) Polyurethane filter pad.

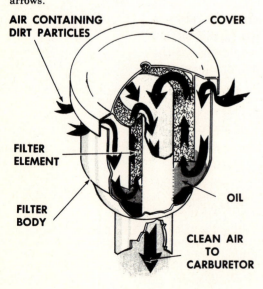

FIGURE 11.5 Oil-bath air cleaner for a small engine. Air flow is shown by arrows.

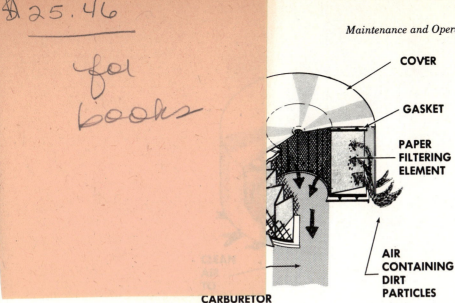

COVER

GASKET

PAPER
FILTERING
ELEMENT

AIR
CONTAINING
DIRT
PARTICLES

CARBURETOR

FIGURE 11.7 Dry-filter air cleaner. The accordian-folded paper filter element is sealed at top and bottom by gaskets so that all entering air must pass through the pores of the paper. The pores trap dirt and dust particles.

oil as a fine mist which is carried up to the metal mesh filter element. Dust particles are trapped by the oily surfaces and the oil washes them down into the oil cup. In the oiled-filter cleaner, shown in Fig. 11.6, the filter element consists of a metal mesh or a polyurethane pad, soaked in oil. The air must pass through the mesh or pad, and dirt particles are trapped by the oily surfaces they pass.

The dry-filter cleaner, shown in Fig. 11.7, has a paper-type or fiber filter element through which the air must pass. The paper is of a very special type, with extremely small pores or openings through which the air can pass. These tiny holes will not, however, permit dust particles to pass so they are trapped on the surface of the filter element.

Regardless of the type of air cleaner, the usual recommendation is that the air cleaners be serviced every 25 hr of service under ideal conditions. If the engine is operated under extremely dirty or dusty conditions, then the air cleaner should be serviced much oftener — as many as two or three times a day! Procedures for cleaning the three types of air cleaners follow.

11.8 Oil-bath Air Cleaner. Before removing the cleaner, disconnect the spark plug wire to make sure the engine will not start. The cleaner will be secured by either a bail wire, or by screw threads on the filter element itself. An oil-bath air cleaner held in place on the carburetor by a wing nut is shown in Fig. 11.8. The air cleaner can be removed by taking off the bail wire or wing nut or unscrewing the

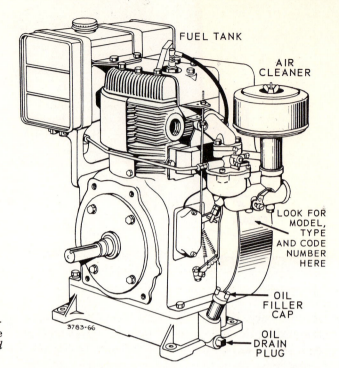

FUEL TANK

AIR CLEANER

LOOK FOR MODEL, TYPE AND CODE NUMBER HERE

OIL FILLER CAP

OIL DRAIN PLUG

3783-66

FIGURE 11.8 Small engine with oil-bath type air cleaner. (*Briggs and Stratton Corporation.*)

filter element, as shown in Fig. 11.9. If there is any danger of dirt or dust falling into the carburetor, cover the opening with a clean cloth or plastic film.

After the air cleaner parts have been removed, separate them and pour out the old oil. Clean the oil cup, filter, and cap with solvent and a brush, as shown in Fig. 11.10. Be sure to remove any caked dirt in the bottom of the oil cup. Refill the cup to the oil-level mark with the oil specified—usually SAE 30—as shown in Fig. 11.9.

Examine the condition of the gasket. If it is at all damaged, it should be replaced. Then, reinstall the oil cup and filter element.

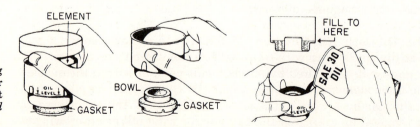

ELEMENT

OIL LEVEL

GASKET

BOWL

GASKET

FILL TO HERE

SAE 30 OIL

OIL LEVEL

FIGURE 11.9 Removing oil-bath air cleaner for cleaning and replacement of the oil. (*Briggs and Stratton Corporation.*)

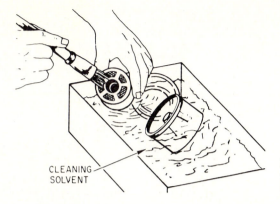

CLEANING
SOLVENT

FIGURE 11.10   Cleaning oil-bath air cleaner in solvent.

11.9 Oiled-filter Air Cleaner.   These are held in place by a snap-on cover, a wing nut, or a screw. Typical oiled-filter air cleaners are shown in Fig. 11.11. Take off the air cleaner and filter element and wash the filter element in kerosene or detergent and water. Cover the carburetor intake. The polyurethane filter is washed by squeezing repeatedly in the kerosene or water to get out all the old oil and dirt. Then wrap it in cloth and squeeze it dry. Finally, saturate it with engine oil and squeeze it to remove the excess.

The metal mesh element is cleaned by washing it in solvent, as shown in Fig. 11.10. It is dried by blowing compressed air through it or swishing it in the air several times, as shown in Fig. 11.12. It should then be dipped in engine oil to reoil it.

Reinstall the filter element and air cleaner assembly. Some polyurethane elements have a coarse filter on the outside and a fine filter in the inside. When installing this type, make sure that the coarse part faces out.

11.10 Dry-filter Air Cleaner.   Remove the filter element cover if it is a separate part. Some air cleaners of this type are made in one piece without a separate cover. Cover the carburetor intake with a cloth or plastic film. The paper filter element can be cleaned by tapping it lightly on a flat surface, as shown in Fig. 11.13. Do not wash it unless the manufacturer's instructions so specify. Getting the paper wet will clog the paper pores and ruin the element. If the dust does not drop off easily or the element is damaged in any way, throw it away and use a new element. Remember that even one pin hole in the element can let enough dust in to ruin the engine in a short time.

If the element is fiber or moss, clean it by blowing compressed air through it from the inside, as shown in Fig. 11.14. Wash it in soap and

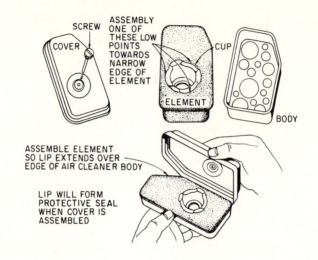

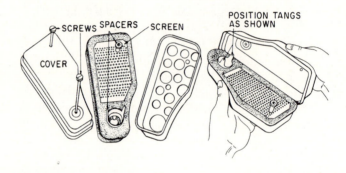

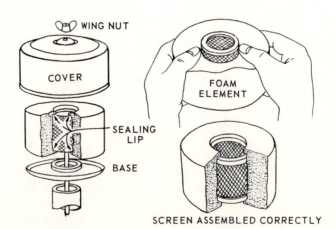

FIGURE 11.11 Typical oiled-filter air cleaners, showing how to remove the filter element from the assembly. (*Briggs and Stratton Corporation.*)

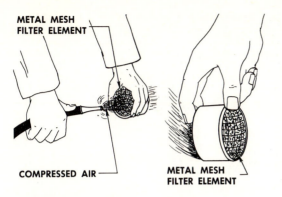
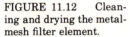

FIGURE 11.12 Cleaning and drying the metal-mesh filter element.

FIGURE 11.13 Tapping the paper filter to knock dirt loose.

water. Do not use an oily solvent because this could clog the element and prevent air from passing through.

On any type of cleaner, be sure that the gasket on which it mounts is in good condition and provides a good seal between the air cleaner and the carburetor intake. A poor seal here will allow unfiltered air to enter the engine, and this means that dirt is getting into the engine. It does not take very much dirt to ruin an engine.

11.11 Cleaning the Fuel Strainer or Filter. There are three general types of fuel strainers—also called fuel filters—the type with a sediment bowl, shown in Figs. 11.15 and 11.16, the type mounted in the fuel tank, as shown in Fig. 11.17, and the type having a weighted strainer at the end of a flexible hose, as shown in Fig. 11.18. The latter type, shown in Fig. 11.18, is used on engines that must operate in any

FIGURE 11.14 Using compressed air to clean the fiber-type filter element. Note that the compressed air is blown from inside out, or in the direction opposite to the air flow during air-cleaner operation.

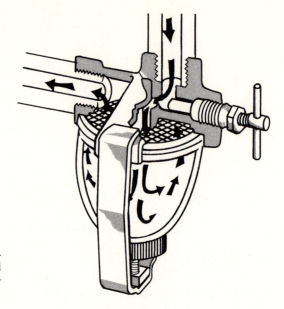

FIGURE 11.15   Sediment-bowl type of fuel strainer. Arrows show direction of fuel flow.

position such as chain saws. Regardless of the position, the weighted strainer will fall to the low part of the tank so the end of the hose will remain in the fuel.

FIGURE 11.16   Sediment-bowl type of fuel filter with the filter bowl removed.

FIGURE 11.17   Fuel strainer of type mounted inside the fuel tank.

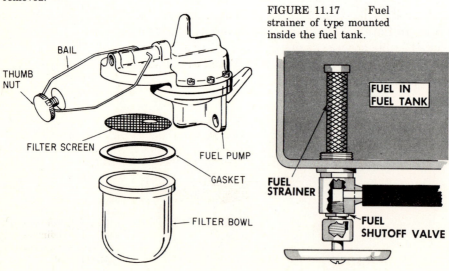

BAIL

THUMB NUT

FILTER SCREEN

FUEL PUMP

GASKET

FILTER BOWL

FUEL IN FUEL TANK

FUEL STRAINER

FUEL SHUTOFF VALVE

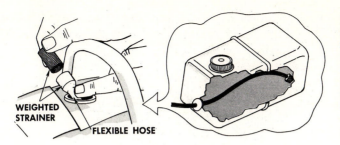

FIGURE 11.18 Type of fuel strainer attached to flexible hose inside the fuel tank. Left, flexible hose partly pulled from fuel tank so strainer can be removed for cleaning. Right, tank partly cut away so hose and strainer can be seen in tank.

11.12 Sediment-bowl Type. To clean the sediment-bowl type fuel strainer, shown in Figs. 11.15 and 11.16, close the shutoff valve on top of the assembly to prevent fuel from flowing out of the fuel tank. Loosen the thumb nut on a wire bail and swing the wire bail to one side. Remove the bowl with a twisting motion. Twisting the bowl reduces the chances of breaking the cork gasket. Remove the gasket and the strainer screen. See Fig. 11.19. The screen is usually held in place by the clamping action between the bowl and gasket. On some, it is held in place by a retainer clip, as shown. Wash the filter screen and dry it. Wash out the sediment bowl and make sure it is clean. Open the shutoff valve and allow about a cupful of gasoline to drain out into a container. This will remove any dirt in the line between the tank and filter. If the fuel flows out very slowly, chances are the vent in the fuel-tank cap is clogged. Remove it to see if the fuel flows more freely. If it does, then clean the cap vents by soaking the cap in solvent.

Install the filter screen, gasket, and sediment bowl. Use a new gasket if needed. If you don't have a new gasket available, turn the old gasket over on reinstallation to get a better seal. Before tightening the thumb nut, open the shutoff valve to fill the sediment bowl. This eliminates any air that might otherwise cause an air lock in the line.

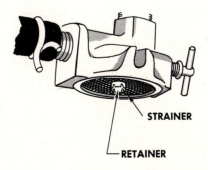

STRAINER

RETAINER

FIGURE 11.19 Upper part of sediment-bowl type of fuel strainer with bowl removed.

11.13 Fuel-tank-mounted Fuel Strainer.   This type of strainer, shown in Fig. 11.17, may or may not be removable from the fuel tank. If it is removable, it can be taken out by unscrewing the fuel shutoff valve or tank fitting on which the strainer is mounted. Clean the strainer by swishing it in solvent and drying it with compressed air. On the type which is permanently mounted in the fuel tank, remove the fuel tank and wash it out with solvent several times to clean the tank and strainer.

11.14 Weighted-strainer Type.   On this type, shown in Fig. 11.18, fish the strainer out of the tank with a bent wire. Remove the strainer from the end of the weighted hose and clean it by swishing it in solvent and then drying it with compressed air.

11.15 Cleaning Crankcase Breathers.   Four cycle engines must have some means of allowing blowby to escape. Blowby, as you know, is the seepage of combustion gases from the combustion chamber past the piston and rings. This blowby can build up in the crankcase if it has no way to escape, and this can damage the engine. The blowby gases can cause corrosion of engine parts and short engine life. To allow the blowby gases to escape, various types of crankcase breathers are used. One type, shown in Fig. 11.20, consists of a mesh-type filter element and a reed valve. The reed valve is a flexible plate which rests against one or more openings. Figure 11.21 is a disassembled view of the arrangement. The reed-valve type operates as follows.

When the piston is moving down on either the power or intake stroke, pressure is created in the crankcase. This pressure pushes the reed valve open so that the blowby gases are forced out of the crank-case, as shown to the left in Fig. 11.20. Then, when the piston moves up, either on the exhaust or the compression stroke, a vacuum is produced in the crankcase. This permits atmospheric pressure to push fresh air into the crankcase. Actually, the reed valve is so arranged that it causes a slight vacuum to be retained in the crankcase. This helps prevent oil leakage through the oil seals and gaskets. The reed valve either has a small hole in it or does not close quite completely; either arrangement puts some restriction on air entering the crank-case so that a slight vacuum remains at the end of the exhaust or compression stroke.

NOTE: Leakage through an oil seal is a pretty good indication of a clogged crankcase breather. If the crankcase breather becomes clogged, excessive pressure will build up in the crankcase, and this can cause oil seals to rupture.

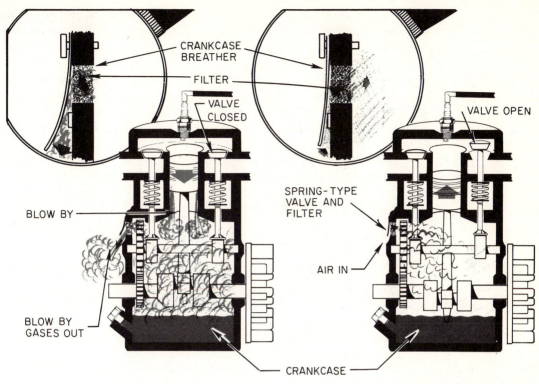

CRANKCASE
BREATHER

FILTER

VALVE
CLOSED

VALVE OPEN

SPRING-TYPE
VALVE AND
FILTER

BLOW BY

AIR IN

BLOW BY
GASES OUT

CRANKCASE

FIGURE 11.20 Crank-
case breather for a small
four-cycle engine. Left,
when the piston is moving
down, pressure in the
crankcase forces the reed
valve to open so blowby
gases are forced out
of the crankcase. Right,
when the piston moves
up, the partial vacuum
created in the crankcase
causes the reed valve to
close.

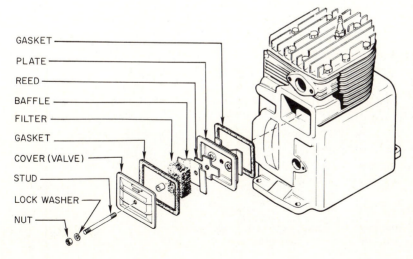

GASKET

PLATE

REED

BAFFLE

FILTER

GASKET

COVER (VALVE)

STUD

LOCK WASHER

NUT

FIGURE 11.21 Disas-
sembled view of a crank-
case breather for a small
four-cycle engine.

FIGURE 11.22   Ball-check type of crankcase breather for a small four-cycle engine.

Other types of crankcase breathers include a ball-check and a floating disk, as shown in Figs. 11.22 and 11.23. They are opened by pressure in the crankcase and partly closed by gravity and by atmospheric pressure.

The filter elements are pads of either metal mesh, fiber, or polyurethane.

To clean a crankcase-breather filter, remove the nut, screw, or other fastener holding it in position. Be sure to note the proper relationships of the various parts. Figure 11.21 shows one arrangement. The filter element can be cleaned in solvent and dried. Reassemble the parts in their original arrangement.

NOTE: Some breathers have oil-drain holes to allow any oil trying to exit from the crankcase to flow back into the crankcase. On these, be sure to install the drain hole toward the base of the engine to permit drainage. If it is installed up, it could cause oil to be pumped out through the breather.

11.16 Changing Oil in Four-Cycle Engines.   Use the type and grade of oil recommended by the engine manufacturer. Oil level in the crankcase should be checked every few hours. Some engines have a dipstick which is attached to the filler plug or cap. On these, the dipstick should be removed, wiped, reinserted, removed again, and the level of the oil on the dipstick noted. On other engines, oil level in the crankcase is correct if the oil is filled to overflowing when the engine is level. See Fig. 11.24. Some engines have an oil-minder, as shown in Fig. 11.25. To use it, you do not have to remove the oil-filler plug. Instead, you simply press the bellows down several times, as

FIGURE 11.23   Floating disk type of crankcase breather for a small four-cycle engine.

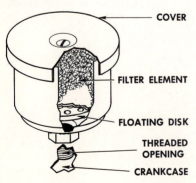

FIGURE 11.24   Oil-filler plug on side of crankcase. Oil level is correct when the crankcase is filled to overflowing when the engine is level. (*Briggs and Stratton Corporation.*)

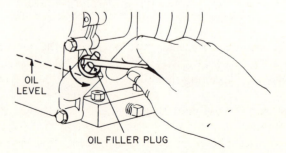

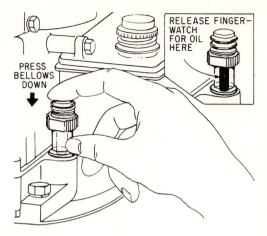

FIGURE 11.25   Oil minder which permits check of oil level in crankcase without removing the oil-filler plug. Pressing the bellows down several times will draw oil up into the plastic tube if the oil level is high enough. (*Briggs and Stratton Corporation.*)

shown, and note whether or not the plastic tube fills with oil. If it does not, the oil level is low.

NOTE: Always make sure to remove all dirt from around the filler plug before removing it so that no dirt will fall into the crankcase.

If oil is needed, fill the crankcase to overflowing or until it is at the "full" mark on the dipstick.

Oil should be changed periodically as specified by the engine manufacturer. The time interval usually specified is 25 hr for normal conditions or oftener for dusty conditions. Many machines have a drain plug at the bottom of the oil sump, as shown in Fig. 11.26. The engine must be turned on its side so the plug is at the bottom, as shown, before the plug is removed. Or, if you prefer, the oil can be drained through the oil-filter opening, as shown in Fig. 11.27, with the engine turned on its side.

NOTE: It is best to drain the oil with the engine hot. This is added insurance that all the sludge and dirt in the bottom of the crankcase will drain out. If the engine is cold, this sludge tends to harden and is slow to drain.

After the drainage is complete, replace the oil-drain plug if it has been removed, set the engine upright, and fill the crankcase to the proper level with the recommended oil. Replace the filler plug or cap.

11.17 Lubricating Two-Cycle Engines.   As you know, lubrication of two-cycle engines is provided by adding oil in the recommended amount to the fuel. The oil-fuel mixture enters the crankcase, as

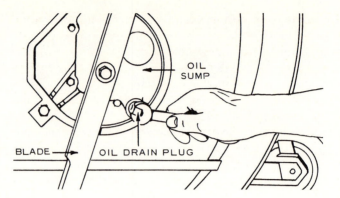

FIGURE 11.26   Removing drain plug to drain oil from crankcase. (*Briggs and Stratton Corporation.*)

explained previously. The fuel, in vapor form, passes on up to the combustion chamber as a component of the air-fuel mixture. Part of the oil, in mist form, is retained in the crankcase where it lubricates the piston, rings, and crankshaft bearings. Some of the oil does get into the combustion chamber where it is burned along with the air-fuel mixture.

The amount of oil to be mixed with the fuel is critical, and the manufacturer's recommendations should be carefully followed. Adding too much oil will cause the exhaust ports to become clogged very quickly, and carbon deposits will form on the piston and rings. This causes poor engine performance. Adding too little oil will deprive the engine of adequate lubrication, so it will wear out that much sooner.

To mix oil and fuel, use an approved metal gasoline can. Never store gasoline in glass containers, such as gallon jugs. This is very dangerous. If you should happen to break a glass container, gasoline would flood the area and could ignite. This might result in a serious or even fatal fire. Don't take chances; use an approved metal container.

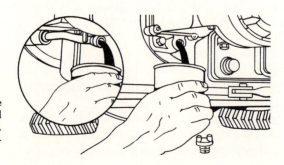

FIGURE 11.27   Engine tilted on side so old oil can drain from crankcase. (*Briggs and Stratton Corporation.*)

Some containers have an oil-measuring cup at the filler opening. This makes it easy to measure out the correct amount of oil to be added when the container is to be refilled. First, add the gasoline to fill the container about half full. Then, measure out and add the oil. Close the container and shake it vigorously. Add the rest of the fuel and shake the container again, very vigorously. Make sure that the oil and fuel are thoroughly mixed.

When storing the oil-fuel mixture, be sure to mark the container so the mixture will not be used in a four-cycle engine. The oil-fuel mixture in a four-cycle engine will cause trouble—heavy carbon formations on the piston, and valves, rings, and a fouled spark plug from carbon.

> CAUTION. **Always turn the engine off before adding gasoline. Never add gasoline to a tank when the engine is running. It could slop over and catch on fire. The flames could cause the gasoline container from which you are pouring to explode. For the same reason, do not smoke or have any flame going near you when you are refueling the engine.**

11.18 Storing Gasoline. You will find that there are local and state laws about storing gasoline. These laws are for your own protection and they should be carefully obeyed. A basic rule is to never store gasoline in a closed room where gasoline vapors will accumulate. Terrible explosions have resulted from storage of gasoline in a container that was not tightly sealed, in a closed room. The accumulating gasoline vapors can be ignited by the spark from turning a light switch on or off, or from a spark caused by one metal object striking another.

Gasoline that is stored for any length of time deteriorates. The length of time that the gasoline stays good depends on its composition and the additives that have been put into it. This is the reason that any machine that is to be stored for any length of time should have its fuel tank and carburetor drained of all fuel. Otherwise, the stale gasoline can deposit gum and varnish on critical parts. For instance, carburetor jets can become clogged, and this will cause poor engine performance and require a complete carburetor overhaul to set things right again. For the same reason, you should not store gasoline in containers for long periods and then expect to use it.

11.19 Starting, Operating, and Stopping Engine. There are more complaints about hard starting than about anything else on small engines. You can start small engines more easily if you will follow the correct procedure. Further, proper operation and stopping of the engine will prolong its life. Here are some hints.

1. Starting the Engine.   If the engine uses a rope-wind starter, make sure the equipment is level. This reduces the possibility of tipping the equipment over when you pull the rope. You can guard against this by putting one foot on the equipment—in this case a lawn mower—as shown in Fig. 11.28.

> CAUTION.   **Never put your foot under the lawn mower where the rotating blade might hit it. You could be seriously injured. In fact, this is good advice when working around any machinery. Stay clear of moving parts!**

Never start and operate the engine in a closed place, such as a garage with the doors closed. The engine can produce enough carbon monoxide in a few minutes to kill you!

If the equipment has brakes, apply them when starting the engine. If it has a clutch, disengage it if possible so the machine will not start to move when the engine starts.

Many engines have a shutoff valve between the fuel tank and the carburetor. If it has been turned off, turn it on again.

Close the choke valve or prime the engine. We have described chokes and primers in an earlier chapter. Some manufacturers recommend that you turn the engine over a few times with the ignition off to allow gasoline to get up into the carburetor. Then, when you crank with the ignition on, the start will be easier and quicker.

On riding equipment, operate the controls from the driver's seat. Then, if anything goes wrong, or the equipment suddenly takes off, you can quickly stop the engine.

Adjust the throttle to the recommended opening for starting. Some engines have a single control for choke and throttle. On these, move the lever to the choke position. Then, when the engine starts, move it back to the open throttle position.

FIGURE 11.28   Make sure the mower is level and that you have it under control by holding it or by having a foot on it before using the rope-wind starter.

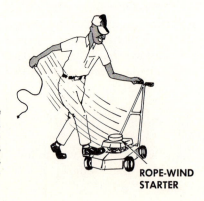

ROPE-WIND STARTER

Turn on the ignition and crank the engine. If you are using a rope-wind or rope-rewind starter, pull the rope until the engine reaches the compression stroke. Then rewind the rope so you can give it a good hard pull through the compression stroke.

> CAUTION. **When starting a chain saw, put it on the ground or brace it so it will not get out of control when you crank it. If you don't have full control, the saw could get away from you with disastrous results such as a badly cut leg or arm.**

If you are using an electric starter, close the starter switch and allow the engine to crank until it starts or until you have cranked for 10 sec. Avoid long cranking periods because this can damage the starter.

If the engine does not start right away, open the choke valve part way and try again. Chances are the engine has flooded; that is, it has gotten too much gasoline.

Once the engine has started, allow it to operate at fast idle for a minute or two so that the engine has a chance to warm up. Never gun a cold engine or try to take full power from it. Give it a chance to warm up first.

If you have trouble starting, refer to the trouble-diagnosis procedure outlined in Chap. 10.

2. Operating the Engine.   Overloading and overspeeding are the two most common causes of small-engine trouble and short engine life. Overspeeding the engine by improperly adjusting the governor will shorten engine life and can actually cause the engine to blow up from the excessive speed. Also, high speed can spin the operating parts of the equipment faster than designed speed, with very damaging results. For instance, the tips of lawn mower blades should never exceed 19,000 ft/min (feet per minute), and manufacturers design their equipment to hold engine speed down so this upper limit is never exceeded. If the engine is overspeeded sufficiently, the blade might explode and parts would fly off and could seriously injure someone.

Overloading an engine can cause the engine to overheat so that engine parts will wear rapidly. If you have to use the equipment in heavy going—for instance cutting tough, tall, wet grass—take it easy. Cut a narrow swath, and move the mower slowly.

If the engine is new or has just been rebuilt, give it a break and allow it to work up to its maximum potential gradually. On two-cycle engines, adjust the carburetor for a fairly rich mixture for the first 10 hr. On four-cycle engines, adjust the throttle to medium speed and allow the engine to operate for about 30 min. without load. Change the oil on four-cycle engines after the first few hours of operation. Follow the instructions on the nameplate attached to the engine or equipment.

> CAUTION.   **When working on the "business" side of equipment – under the mower where the blade is, the chain on the chain saw, etc. – always make sure the engine cannot start by turning the engine off and by disconnecting the wire from the spark plug.**

3. Stopping the Engine.   Although you might think there is nothing special about stopping an engine, there are actually right and wrong ways to do this. First, remove any load from the engine before turning it off. Then reduce engine speed to idle and allow it to run for a minute or two before turning it off. This cools the engine down more gradually and reduces thermal stresses on engine parts.

After turning the engine off, close the fuel-tank shutoff valve if the engine has one. This takes any pressure off the carburetor diaphragms or float and prevents fuel leaks.

If you do not plan to use the engine for a month or so, then drain the fuel tank and carburetor to prevent formation of gum which could clog the carburetor passages. If you are not going to use the engine for a longer period, treat it as explained in Sec. 11.20.

11.20 Winter Storage.   For winter storage, drain the fuel tank, and run the engine to use up the fuel in the carburetor. Fuel left in the carburetor is apt to form gum that will clog fuel passages. Remove the spark plug and pour a tablespoon or so of heavy engine oil into the combustion chamber. Turn the engine over a few times to distribute the oil over the engine parts. Replace the plug. Store the machine in a warm, dry place.

> CAUTION.   **As we have said previously, when you are working on the "business" side of equipment (under the mower where the blade is, on the chain saw, etc.) always make sure the engine cannot start by turning the engine off and by disconnecting the spark plug wire.**

CHECKUP

You have now covered the fundamentals of engine maintenance and operation. That is, in the chapter you have just completed, you have learned about the things you should do to keep the engine in good operating condition, the proper way to operate the engine, and how to store it when it is to be idle for a period of time. The following questions will not only give you a chance to check up on how well you understand and remember these fundamentals, but also will help you to remember them better. The act of writing down the answers to the questions will fix the facts more firmly in your mind.

NOTE: Write down your answers in your notebook. Then later you will find your notebook filled with valuable information which you can refer to quickly.

Completing the Sentences: Test 11. The sentences below are not complete. After each sentence there are several words or phrases, only one of which will correctly complete the sentence. Write each sentence in your notebook, selecting the proper word or phrase to complete it correctly.

1. One of the most important maintenance jobs is to (*a*) operate the engine frequently; (*b*) keep the engine clean; (*c*) keep the fuel tank full.
2. If you are asked whether or not you should operate an engine without the shrouds and baffles, you should answer (*a*) no, never; (*b*) yes, at times; (*c*) only when cleaning the engine.
3. Trash on the cylinder and head fins can cause (*a*) high fuel consumption; (*b*) engine overspeeding; (*c*) engine overheating.
4. To clean carbon from the exhaust ports, the piston should be (*a*) at BDC; (*b*) removed; (*c*) near TDC.
5. The three types of carburetor air cleaners for small engines are (*a*) oil-bath, oiled, dry; (*b*) fiber, paper, mesh; (*c*) mesh, fiber, oiled.
6. To allow blowby to escape from the crankcase, small four-cycle engines have (*a*) fuel strainers; (*b*) blowby valves; (*c*) crankcase breathers.
7. Leakage of oil through a crankshaft seal of a four-cycle engine often means (*a*) a clogged crankcase breather; (*b*) worn bearings; (*c*) loss of compression.
8. The most common complaint about small engines is probably (*a*) high gasoline consumption; (*b*) noise; (*c*) hard starting.

Written Checkup

In the following, you are asked to write down, in your notebook, the answers to the questions asked or to define certain terms. Writing the answers down will help you to remember them.

1. Make a list of the maintenance hints which will help to keep a small engine healthy.
2. Explain the procedure for cleaning a small engine. Why is this important?
3. Why must you never operate an air-cooled engine with the shrouds and baffles removed?
4. Explain how to clean the exhaust ports in a two-cycle engine. What happens if these ports become clogged with carbon?

5. List the various kinds of carburetor air filters used on small engines and explain how to clean each type.
6. List the different kinds of fuel filters used on small engines and explain how to clean each.
7. What is the purpose of the crankcase breather in the four-cycle engine?
8. Explain how to change the oil in a four-cycle engine.
9. What is the purpose of putting oil in the fuel used in a two-cycle engine? How does the oil do its job in the engine?
10. Discuss the basic rules for storing gasoline.
11. Describe the full procedure for starting a small engine.
12. What are the major points to watch when operating a small engine?
13. What are the major things to do when stopping an engine?
14. Explain how to put up a small engine for winter storage.

Servicing Mechanical Starters

12

In this chapter, we will describe the servicing procedures required to service the different kinds of mechanical starters used on small engines. As you will recall, from Chap. 9, there are four general types of small-engine starters: rope-wind, rope-rewind, windup, and electric. The electric starters are of two types, 12-volt and 120-volt. We will look at the rope-wind, rope-rewind, and windup starters in this chapter. In the following chapter, we will discuss servicing of electric starters.

12.1 Rope-wind Starter. The only things requiring service with this starter are the rope and handle. You can buy a new rope and handle assembly, or if the handle is okay, you can install a new rope if the old one goes bad. The new rope should be the same size and length as the old one. Most small engines use a $\frac{3}{16}$-in. (inch) nylon braided rope about 5 ft long. The rope should be long enough to wrap around the flywheel pulley up to five times plus about 1 ft to make the knots at the two ends.

Either of two kinds of knots can be used, and the procedures for making them are illustrated in Figs. 12.1 and 12.2. Before pushing the rope through the handle, singe both ends of the rope with a match to prevent raveling of the rope and to assure that the knot will not slip and loosen. Then insert the rope through the handle and tie the knots in the two ends of the rope.

12.2 Rope-rewind Starters. There are several designs of rope-rewind starters. All work on the same general principle but require different servicing procedures. We will first cover the general servicing procedures, and then examine several specific models.

12.3 General Servicing Procedures on Rope-rewind Starters. General servicing procedures for rope-rewind starters include replacing the rope, replacing the spring, and repairing the drive mechanism. If you have any question in your mind about how this type of starter works, refer to Sec. 9.2.

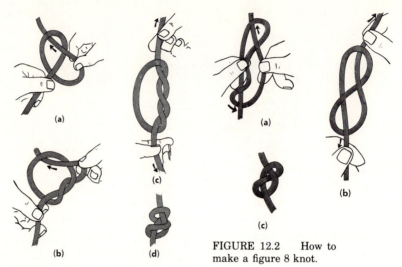

(a)

(b)

(c)

(d)

FIGURE 12.1   How to make a double-overhand knot.

(a)

(b)

(c)

FIGURE 12.2   How to make a figure 8 knot.

1. Replacing Rope. Remove the starter from the engine. Usually, it is attached by three or four **screws**. On some engines, it will be necessary to remove a shroud or piece of equipment to get to the starter. You can work on the starter if it is laid on a workbench, but you will find it easier to work on if you clamp it in a vise, as shown in Fig. 12.3. If the knot in the rope is visible, you will be able to replace the rope without disassembling the starter. If the knot is not visible, you will have to disassemble the starter.

a. Visible-knot Starter. On the visible-knot starter, shown in Fig. 12.3, pull the rope all the way out. The pulley must be held so the spring will not unwind when the new rope is installed. You can hold the pulley with a pair of vise-grip pliers or with a wrench and square stock, as shown in Fig. 12.4. Note that the end of the wrench is wired to the housing, as shown in the insert.

If the spring is unwound, you can wind it up again with a wrench and square stock, as shown in Fig. 12.4, or with a screwdriver, as shown in Fig. 12.5. The general rule is to wind the spring up tight, and then back it off one complete turn.

The new rope should be the same size and length as the old. Singe both ends of the rope with a match flame. This prevents the rope from fraying and also prevents the knot from loosening. Thread the rope

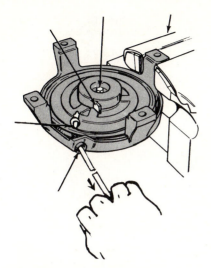

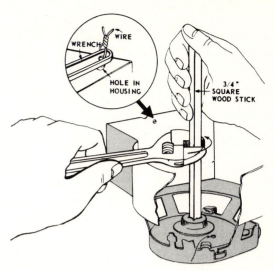

FIGURE 12.3   To begin
the repair of the visible-
knot type rope-rewind
starter, pull the rope all
the way out.

FIGURE 12.4   You can
hold the pulley, or wind
up the recoil spring, with
a square piece of wood and
a wrench on some models.

through the housing and hole in the pulley, as shown in Figs. 12.6
and 12.7. Using a hooked wire, as shown in Fig. 12.6, or a special rope
inserter made of spring wire, as shown in Fig. 12.8, will make the job
easier on the design shown in Fig. 12.6. Tie a knot in the end of the

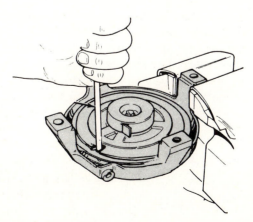

FIGURE 12.5   On other
models, you can wind up
the recoil spring with a
screwdriver.

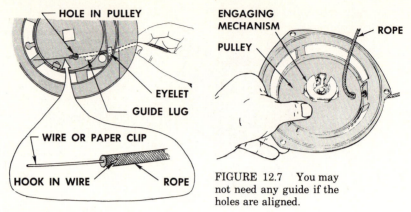

FIGURE 12.6 When re-threading the rope into the pulley hole, you may find that a piece of wire hooked into the end of the rope, as shown at the bottom, will make it easier.

FIGURE 12.7 You may not need any guide if the holes are aligned.

rope (see Figs. 12.1 and 12.2) and push the knot down into the hole in the pulley where it will not interfere with anything when the pulley rotates.

b. Hidden-knot Starter. If the knot is not visible, the starter must be disassembled to replace the rope. Disconnect the handle from the end of the rope and untie the knot. Hold the pulley with a cloth or a gloved hand and allow it to turn slowly. This removes the tension on the spring. Then remove the starter drive from the pulley assembly. This requires disassembly of the starter-drive mechanism which we will cover on a following page. There are two types of pulley, the one-piece and the two-piece. If the starter has a one-piece pulley, remove the pulley, leaving the spring in the housing. If it has a two-piece pulley, remove one side of the pulley, leaving the spring under the other half in the housing. You will now be able to remove the rope and install a new rope, as we have already explained.

FIGURE 12.8 Rope inserter made of a heavy wire, flattened at one end, and stuck in a wood handle, as shown. (*Briggs and Stratton Corporation.*)

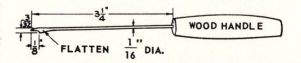

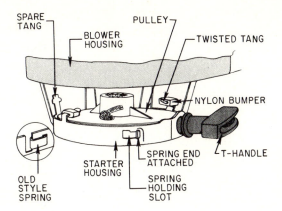

FIGURE 12.9   You recognize this as a removable type of spring because you can see the spring end from outside.

You now reinstall the pulley after winding the rope on it and install the drive mechanism as we will explain later. Rewind the spring and secure it with vise-grip pliers or with a wrench, as shown in Figs. 12.4 and 12.5. Reinstall the starter on the engine, making sure that it is properly aligned. Some starters have an alignment rod which is inserted into a hole in the crankshaft to secure alignment. If the alignment is not correct, the starter will not work properly and will wear out rapidly.

2. Replacing Spring.   There are two types of recoil springs, the removable type and the packaged type. One removable type is shown installed in a starter in Fig. 12.9. You can recognize it because you can see the spring end on the outside. The packaged type is a stronger spring and comes preoiled and compressed inside a retainer housing. The retainer housing may be of the permanent type and is installed in the starter along with the spring. On others, the housing is discarded when the spring is installed in the starter.

If the spring is of the removable type accessible from outside the starter housing, remove the starter and clamp it in a vise. Remove the rope and release the spring tension, as already explained. Use pliers to pull the spring out as far as possible (Fig. 12.10). Use a cloth or wear gloves to protect your hands. Remove the pulley. If the pulley is held in place by tangs, as shown in Fig. 12.9, bend up one tang to remove the pulley. If the pulley is held in place by the starter-drive mechanism, remove it first. We will describe a starter-drive service later.

Disconnect the spring from the pulley, as shown in Fig. 12.11, and if you are going to use it again, straighten it, as shown in Fig. 12.12.

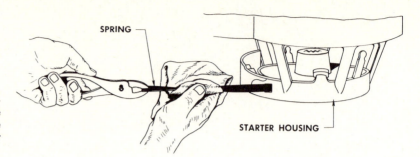

FIGURE 12.10   Use pliers and a cloth or a glove on your hand to pull the spring out of the starter housing.

Wear gloves to protect your hands. Attach the spring to the pulley and install the pulley. Wind up the spring as already explained (see Figs. 12.4 and 12.5). Attach the rope to the pulley. Install the starter on the engine and check its operation.

If the spring is of the packaged type, as shown in Fig. 12.13, the retainer housing may be of the temporary type or it may be permanent. If temporary, the spring is merely slid from the temporary housing into the permanent housing on the starter. If it is of the permanent type, the old housing is removed from the starter and the new housing, with spring, is installed.

If the spring is of the semicoiled type, it should be removed from its package, as shown in Fig. 12.14. It can then be wound into the housing on the starter.

FIGURE 12.12   Straightening the spring to provide more tension. Wear gloves when handling the spring.

FIGURE 12.11   Unhooking the spring from the pulley.

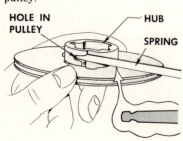

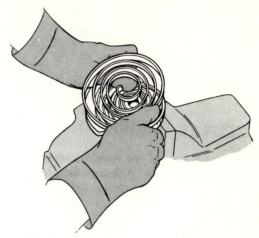

FIGURE 12.13   Most replacement recoil springs are enclosed in a housing for safe handling.

FIGURE 12.14   Some recoil springs are partly coiled but not packaged.

3. Drive Mechanism Service.   All rope-rewind starters are similar in construction, the main difference being in the ratchet mechanism that causes the pulley to engage the crankshaft for cranking, and then releases when the engine starts. There are four general types, as shown in Figs. 12.15 to 12.18. Study these four illustrations to see how the four types work.

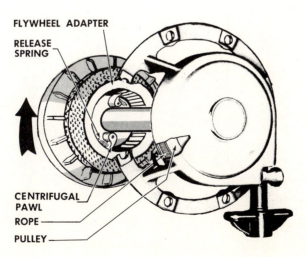

FLYWHEEL ADAPTER

RELEASE SPRING

CENTRIFUGAL PAWL

ROPE

PULLEY

FIGURE 12.15   Rope-rewind drive mechanism using centrifugally actuated pawls.

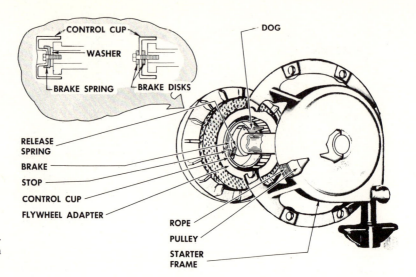

FIGURE 12.16   Rope-rewind drive mechanism using cam-operated dog.

To service the drive mechanism, remove the starter and clamp it in a vice, as shown in Fig. 12.3. Pull the rope and note the action of the drive mechanism. If service is required, remove the drive mechanism, as shown in Fig. 12.19. Disassemble it. Note very carefully what parts go where when you take the mechanism apart so you will know exactly how everything goes together when you reassemble it. Be

FIGURE 12.17   Rope-rewind drive mechanism using cam-operated shoes.

FIGURE 12.18   Rope-rewind drive mechanism using wedging steel balls.

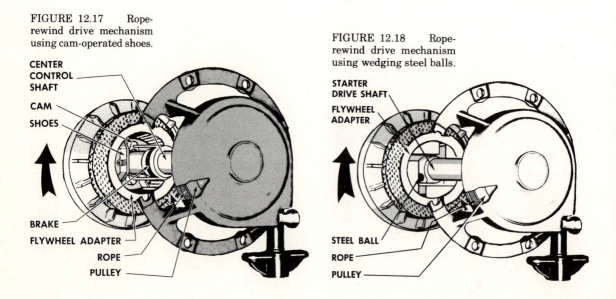

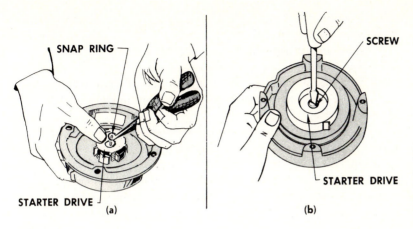

FIGURE 12.19 On some models, you remove the drive mechanism by removing a snap ring as at (*a*). On others, you remove a screw as at (*b*).

especially careful to put the engaging mechanism back in exactly the same way as you found it. If you get it in wrong, the starter will not work. Lubricate the parts lightly with graphite or multipurpose grease on reassembly.

Check for proper operation after reassembly and then install the starter on the engine. Again check it for proper operation.

12.4 Lawn Boy C-10 and C-12 Rope-rewind Starters. These two models are very similar in construction. Figure 12.20 shows the principle of operation. A spring-loaded pin, attached to the flywheel, engages with the starter pulley so the flywheel turns when the pulley is rotated. The pulley rotates when the rope is pulled. This not only

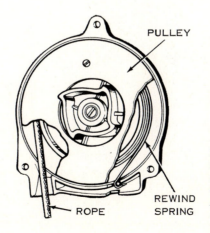

FIGURE 12.20 Rope-rewind starter partly cut away so rewind spring can be seen. (*Lawn Boy Division, Outboard Marine Corporation.*)

spins the flywheel to start the engine, but it also winds up the spring which is attached to the pulley at one end and to the starter frame at the other end. When the engine starts, centrifugal force moves the pin outward so that it is disengaged from the starter pulley. At the same time, releasing the rope allows the spring to turn the pulley back in the reverse direction, thus winding up the rope on the pulley in readiness for another start.

Figure 12.21 shows the C-12 rope-rewind starter in disassembled view. On this model, the rope is locked into the pulley and handle by pressing it into place, as shown in Fig. 12.22. No separate clamp or bead is necessary. On the earlier model (C-10), the rope was retained in the pulley by a metal bead clamped on the end of the rope, as shown in Fig. 12.23, or it was clamped into place with a clamp plate, as shown in Fig. 12.24.

To replace the rope, remove the three screws and lift the starter from the base. Be very careful not to release the spring. It is coiled under tension in the cap. Leave the spring in the cap unless it is damaged and must be replaced. Then remove the three screws holding the plate on the pulley and pry the rope out of the anchor in the pulley.

FIGURE 12.21   Disassembled view of the C-12 rope-rewind starter. (*Lawn Boy Division, Outboard Marine Corporation.*)

FIGURE 12.22   Methods of attaching rope to the handle and pulley. (*Lawn Boy Division, Outboard Marine Corporation.*)

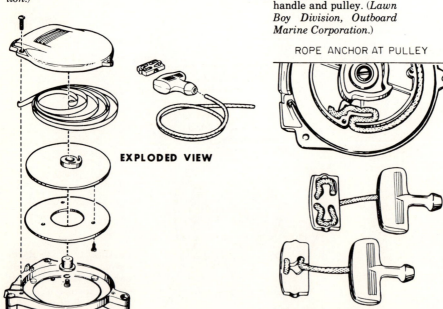

EXPLODED VIEW

ROPE ANCHOR AT PULLEY

ROPE ANCHOR AT HANDLE

See Fig. 12.22. Install the new rope by pressing it into place in the same position, as shown.

To replace the spring in the starter cap after the old spring has been removed, anchor the rope and wind it in the pulley. Put inside loop of the spring into hub of the pulley, as shown in Fig. 12.25. Put pulley on starter cap so that the spring runs through the slot in the cap. Secure the pulley to the starter cap with the pulley bearing, washer, and screw. Tighten the screw securely.

Grip the cap firmly and pull the rope to turn the pulley, as shown in Fig. 12.26. The turning pulley will draw the spring into the cap. Rewind the rope on the pulley and repeat the operation until the spring has been drawn all the way in. Then, with the rope fully wound on the pulley, turn the pulley against the spring tension and, with the rope handle in the slot, put the starter on the base. Hold the starter in place and pull out on the handle to feel for spring tension (Fig. 12.27). There must be some tension or else the handle will not return all the

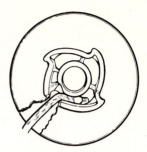

FIGURE 12.23   Metal bead on end of rope is used to retain rope in C-10 model. (*Lawn Boy Division, Outboard Marine Corporation.*)

FIGURE 12.24 On some models of the C-10, the rope is retained by a rope clamp. (*Lawn Boy Division, Outboard Marine Corporation.*)

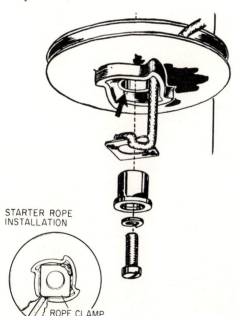

STARTER ROPE INSTALLATION

ROPE CLAMP

FIGURE 12.25 Methods of attaching the inside end of spring to the hub of the pulley. (*Lawn Boy Division, Outboard Marine Corporation.*)

C-12

C-10

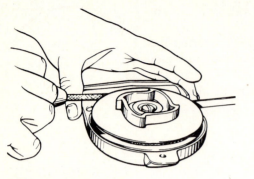

FIGURE 12.26 Turning the pulley by the rope to wind the spring into the cap. (*Lawn Boy Division, Outboard Marine Corporation.*)

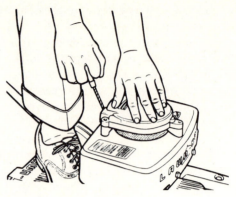

FIGURE 12.27 Testing for proper spring tension before reattaching the starter to the engine. (*Lawn Boy Division, Outboard Marine Corporation.*)

way to the slot. If tension is needed, pull out the rope some more and then rewind it on the pulley, as already explained.

> CAUTION. **Never wind pulley over two turns after tension is felt on the spring.**

Reattach the cap to the base after the correct tension is achieved.

NOTE: Make sure that the pin or pins on the flywheel pulley are on the outside of the starter pulley ratchets before tightening the starter-to-shroud screws.

12.5 Fairbanks-Morse Rope-rewind Starters. This starter, used on some Kohler small engines, is shown disassembled in Fig. 12.28. Figure 12.29 illustrates various steps in servicing this type of rope-rewind starter. If the rope or spring requires replacement, remove the starter from the engine. Use a screwdriver, as shown in Fig. 12.29, hold the washer No. 7 with your thumb, and remove retainer ring No. 6. Then take off washer No. 7, spring No. 8, washers No. 9 and 10. Next, remove the friction shoe assembly including parts No. 11 to 14.

On model 425, relieve tension on the rewind spring by removing the handle and then allowing the rotor to unwind slowly. On model 475, hold the rotor No. 17, as shown in Fig. 12.29, while removing the screws and flanges No. 3 and 5. Then gradually release rotor to allow spring to slowly unwind.

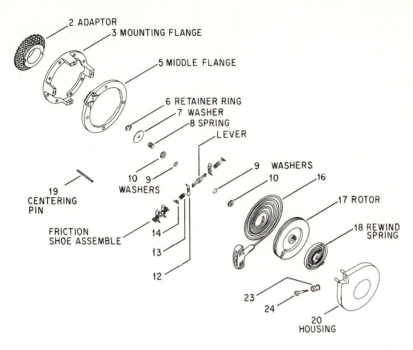

2 ADAPTOR
3 MOUNTING FLANGE
5 MIDDLE FLANGE
6 RETAINER RING
7 WASHER
8 SPRING
LEVER
9 WASHERS
10
10 9
WASHERS
19
CENTERING
PIN
FRICTION
SHOE ASSEMBLE
14
13
12
9
16
17 ROTOR
18 REWIND
SPRING
23
24
20
HOUSING

FIGURE 12.28 Disassembled view of one model of Fairbanks-Morse rope-rewind starter. (*Kohler Company*.)

FIGURE 12.29 Steps in servicing the starter shown in the previous illustration. (*Kohler Company.*)

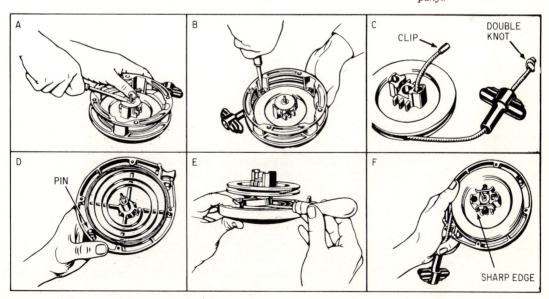

Next, detach spring from rotor by slightly raising the rotor, as shown in Fig. 12.29 and detaching the inner end of the spring from the rotor, using a screwdriver, as shown. This leaves the spring in the cover.

To replace the rope No. 16, thread it through the rotor hole, as shown in Fig. 12.29, and then wind the rope on the rotor. Put the handle on the rope and tie the end of the rope in a double knot, as shown. Singe the ends of the rope, as already explained, before tying the knots.

To replace the spring in the cover, first remove the old spring, turn by turn, holding back the rest of the turns so the spring will not jump out and hurt someone. Note position of the spring loop so when you start to put in the new spring, you can put the loop into the cover in the same position. Replacement springs come with temporary spring holders which prevent the spring from unwinding until it is put into the cover. Put the spring into the cover with the outside loop around the pin, as shown in Fig. 12.29. Then press the spring into the cover and remove the spring holder. Apply a few drops of SAE 20 or 30 oil to the spring, and put some light grease on the cover shaft.

On model 475, put the rotor, complete with cord and handle, into the cover No. 20 and hook the inside loop of the spring to the rotor using a screwdriver. On the model 425, just be sure that the rope is completely wound on the rotor before installing it in the cover.

Pretension the rotor as follows. On model 425, rotate the rotor five turns with the aid of the cord in the cranking direction, as shown in Fig. 12.29. On the model 475, take four additional turns of the rope on the rotor. Thread the cord through the slot and replace the handle.

Replace the other parts removed during disassembly.

To assure proper centering of the starter, pull out the centering pin No. 19 about $\frac{1}{8}$ in. Put the starter on the four screws, making sure the centering pin engages the center hole in the crankshaft and press it into position. Hold the starter with one hand and put the lockwashers and nuts on the screws. Tighten the screws securely.

12.6 Briggs and Stratton Rope-rewind Starters. Figures 12.30 to 12.32 show different rope-rewind starters used on Briggs and Stratton small engines. To replace the spring, first cut the rope at the starter pulley to remove the rope. Then, grasp the outer end of the spring with pliers, as shown in Fig. 12.33, and pull the spring out as far as possible. Next, bend one of the bumper tangs up and lift out the starter pulley so you can disconnect the spring.

To install the spring, clean the spring in solvent and wipe it clean by pulling it through a cloth. Straighten the spring to allow easier

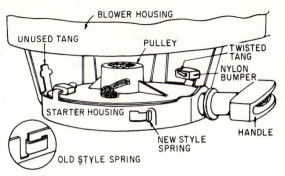

FIGURE 12.30   Typi-
cal rope-rewind starter.
(*Briggs and Stratton Corporation.*)

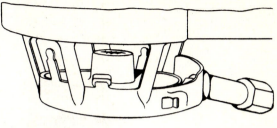

FIGURE 12.31   Rope-
rewind starter models
92500-92900. (*Briggs and
Stratton Corporation.*)

installation and to provide adequate tension. Insert either end of the
spring into the housing slot and hook it into the pulley, as shown in
Fig. 12.34. If the nylon bumpers appear worn, replace them. Put a dab
of grease on the pulley, as shown in Fig. 12.34. Set the pulley into the
housing and bend the bumper tang down, as shown.

Put a $\frac{3}{4}$ in.-square piece of stock into the center of the pulley hub.
Grasp the stock with a wrench and turn the pulley $13\frac{1}{4}$ turns counterclockwise, until the hole in the pulley for the rope knot and eyelet in
the blower housing are in alignment, as shown in Fig. 12.35. Make
sure the spring is securely locked in smaller portion of the tapered
hole, as shown in Fig. 12.36.

FIGURE 12.33   Pulling
spring out with pliers.
(*Briggs and Stratton Corporation.*)

FIGURE 12.32   Rope-
rewind starter model se-
ries 130000, 140000, and
170000. (*Briggs and Stratton Corporation.*)

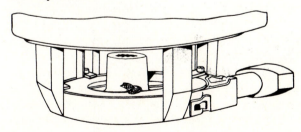

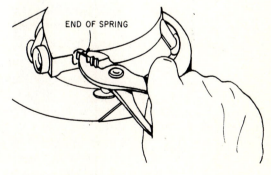

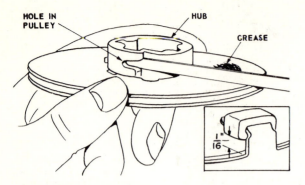

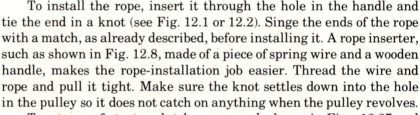

FIGURE 12.34 Install-ing spring. (*Briggs and Stratton Corporation.*)

FIGURE 12.35 Insert-ing rope. (*Briggs and Stratton Corporation.*)

FIGURE 12.36 Locking spring end in housing. (*Briggs and Stratton Cor-poration.*)

To install the rope, insert it through the hole in the handle and tie the end in a knot (see Fig. 12.1 or 12.2). Singe the ends of the rope with a match, as already described, before installing it. A rope inserter, such as shown in Fig. 12.8, made of a piece of spring wire and a wooden handle, makes the rope-installation job easier. Thread the wire and rope and pull it tight. Make sure the knot settles down into the hole in the pulley so it does not catch on anything when the pulley revolves.

Two types of starter clutches are used, shown in Figs. 12.37 and 12.38. When the starter is disassembled, inspect the clutch parts for wear with special attention to the cam areas arrowed in Fig. 12.37. The sealed clutch, shown in Fig. 12.38, can be disassembled by prying the retainer cover from the housing, as shown in Fig. 12.39, using a screwdriver.

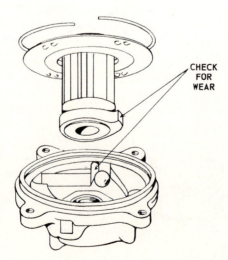

FIGURE 12.37 Ball-type starter clutch which can be disassembled for inspection and cleaning. (*Briggs and Stratton Cor-poration.*)

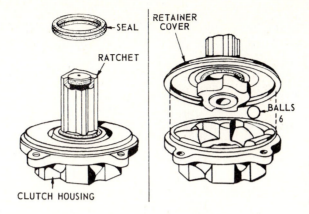

FIGURE 12.38 Sealed-type starter clutches. (*Briggs and Stratton Corporation.*)

12.7 Windup Starter Service. We have already covered the operation of this type of starter in Sec. 9.3. Figure 12.40 shows assembled and disassembled views of windup starters. See also Figs. 9.4 to 9.6. To work on this type of starter, first release the starter control so that the spring unwinds.

> CAUTION. **Never attempt to work on a windup starter without releasing the spring tension. The spring is very strong and could cause serious injury if it should pop out of the starter during disassembly.**

Some windup starters are disassembled from the drive end, others from the handle end. If the handle is riveted or welded to the shaft, remove the drive mechanism first. The drive mechanism and retainer screw are similar to those on rope-rewind starters, but heavier. If necessary, remove the drive mechanism for repair, referring to the story on this in Sec. 12.3. Remove the mainspring assembly, with its retainer.

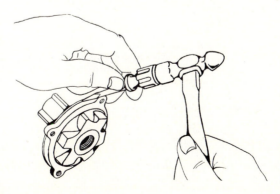

FIGURE 12.39 Disassembling a sealed clutch. (*Briggs and Stratton Corporation.*)

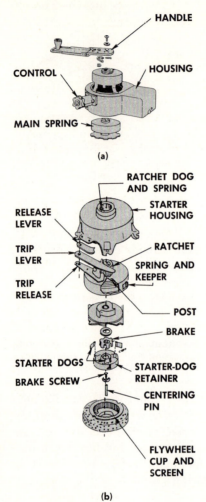

FIGURE 12.40 Assembled and disassembled views of windup starters. In (a), the type shown must be disassembled from the handle end. In (b), the type shown must be disassembled by first removing the drive mechanism.

CAUTION. **Do not remove the spring from the retainer unless the service manual for the engine you are working on specifically says to do it and tells you how. These springs are very strong and can hurt you if they are released!**

When disassembling the drive mechanism, observe carefully the location of all parts so you can put the assembly back together again correctly. Watch for small parts such as springs, washers, and spacers. Check the housing and other operating components for cracks, worn gear teeth or ratchet, and so on.

On reassembly, make sure you put all parts back into their proper places. Reinstall the starter on the engine and check it for proper operation.

NOTE: If you install a new spring, be sure to destroy the old spring so that no one will get hurt tampering with it. The best way is to heat the old spring with a torch. This removes the temper and tension in the spring.

12.8 Briggs and Stratton Windup Starters. These starters use two different kinds of releases. The two types are shown in Figs. 12.41 and 12.42. The control-knob release is used with the unsealed four-ball clutch. The control-lever release is used with the sealed six-ball clutch.

Before working on the starter, be sure the starter spring is released. If it has tension, release it by moving the control knob or lever to the starting position. If this does not release the spring, hold the crank handle with one hand and remove the Phillips-head screw and handle assembly from the starter housing (Fig. 12.43). This will release the spring.

To check for a broken spring with unit still on the engine, put the control knob or lever in the start position and turn the handle 10 turns clockwise. If the engine does not turn over, the spring is broken or else the starter clutch balls are not engaging. Turn the crank handle and watch the starter clutch ratchet. If it does not move the spring is broken.

To disassemble the starter, remove the blower housing and the screw holding the crank handle to the housing. Bend the tangs up

FIGURE 12.41 Old style windup starter with control knob. (*Briggs and Stratton Corporation.*)

FIGURE 12.42   New style windup starter with control lever. (*Briggs and Stratton Corporation.*)

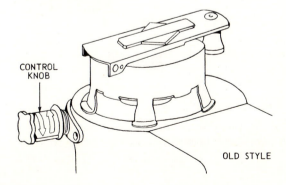

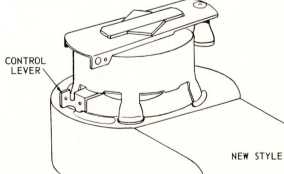

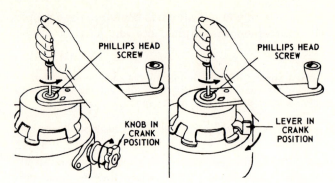

FIGURE 12.43   Releasing spring by removing the Phillips-head screw. Hold the handle with one hand while removing the screw. (*Briggs and Stratton Corporation.*)

that hold the starter spring and housing assembly with a special tool, as shown in Fig. 12.44. Lift the retainer plate, spring and housing assembly out of the blower housing.

CAUTION.   **Do not attempt to take the spring from its housing.**

FIGURE 12.45   Places to apply grease on reassembly. (*Briggs and Stratton Corporation.*)

FIGURE 12.44   Using special tool, shown in detail at the right, to bend the hold-down tangs up. (*Briggs and Stratton Corporation.*)

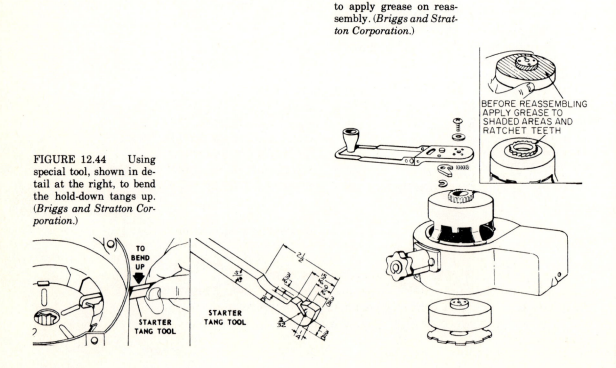

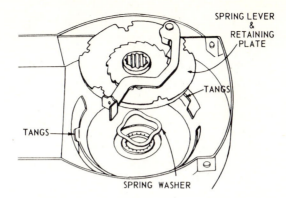

FIGURE 12.46 Spring washer should go into housing, as shown, before putting spring lever and retaining plate in place. Then tangs should be bent down to hold the parts in place. (*Briggs and Stratton Corporation.*)

Check the starter parts for damage. Note the condition of the ratchet gear on the outside of the blower housing for wear. See Fig. 12.45. Do not remove the retaining plate from the spring-and-cup assembly. Check the movement of the control knob or lever to make sure it works easily.

On reassembly, apply grease to the shaded areas and ratchet teeth, as shown in Fig. 12.45. Be sure to install the spring washer in the housing before placing the cup, spring, and release assembly into the housing. Bend the retaining tangs down securely. See Fig. 12.46. Install the starter on the engine, and check it for proper performance.

CHECKUP

In this chapter you have just completed, you have learned how to service the various kinds of mechanical starters used on small engines. The fundamental servicing operations for each type were outlined in this chapter. The following questions will not only give you a chance to check up on how well you understand and remember these fundamentals, but also will help you to remember them better. The act of writing down the answers to the questions will fix the facts more firmly in your mind.

NOTE: Write down your answers in your notebook. Then later you will find your notebook filled with valuable information which you can refer to quickly.

Completing the Sentences: Test 12. The sentences below are not complete. After each sentence there are several words or phrases, only one of which will correctly complete the sentence. Write each sentence in your notebook, selecting the proper word or phrase to complete it correctly.

1. The three general servicing procedures for rope-rewind starters include repairing the drive mechanism and replacing the (*a*) spring or rope; (*b*) spring or crank; (*c*) switch or rotor.
2. There are two types of replacement recoil springs supplied for servicing the rope-rewind starter, the removable type and the (*a*) flat type; (*b*) coiled type; (*c*) packaged type.
3. Comparing the rope-rewind starter and the windup starter, the one that is easiest to service is the (*a*) rope-rewind starter; (*b*) windup starter.
4. The principle of the rope-rewind starter is that a recoil spring (*a*) spins the crankshaft; (*b*) rewinds the rope; (*c*) engages the crank.
5. The principle of the windup starter is that the spring (*a*) spins the crankshaft; (*b*) rewinds the rope; (*c*) engages the crank.

Written Checkup

In the following, you are asked to write down, in your notebook, the answers to the questions asked or to define certain terms. Writing the answers down will help you to remember them.

1. Explain how to make the two kinds of knots described in the chapter.
2. Write down a detailed description of how to service a rope-rewind starter, including tying the knots, replacing the spring, and servicing the drive mechanism.

Electric Starter and Charging-system Service

This chapter covers the servicing of electric starters and charging systems for small engines. There are two general types of electric starters, 12-volt and 120-volt. If the engine has a 12-volt starter, it probably has a charging system, too. The 12-volt starter needs a 12-volt storage battery, and to keep the battery charged, a generator or an alternator is required. The generator or alternator needs a regulator to prevent overcharging of the battery. If a 120-volt alternating-current starter is used, then a charging system is not required although the engine may also have this equipment.

13.1 12-Volt Starter. There are two general types of 12-volt small-engine starters. One uses a Bendix drive, the other a belt drive. Figure 13.1 shows a disassembled view of a Bendix drive starter. Figure 13.2 shows a belt-drive starter mounted on an engine. The belt-drive starter is used with a belt clutch which can drive only one way, from the starter to the engine. When the engine starts and attempts to drive the starter, the clutch disengages.

13.2 12-Volt Starter Troubles. If the starter cranks the engine slowly or not at all, there are four different possible causes to consider:

1. Battery is run down or defective.
2. Starter may be defective.
3. Engine may have internal trouble.
4. Temperature may be too low.

We explain in a following section how to check the battery. If it is in good condition but run down, it will require recharging. Also, the charging system on the engine should be checked to see whether it is doing its job. If the battery is defective, it must be replaced with a new one.

If the battery is in good condition, then the failure to crank properly could be due to some defect in the starter. We will explain how to check out the starter in a following section.

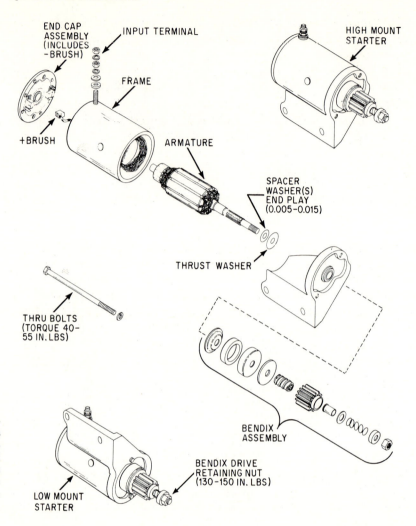

FIGURE 13.1 Disassembled and assembled views of a Bendix drive electric starter. The top and bottom assembled views show two different mounting arrangements. (*Kohler Company.*)

END CAP ASSEMBLY (INCLUDES –BRUSH)

INPUT TERMINAL

HIGH MOUNT STARTER

FRAME

+BRUSH

ARMATURE

SPACER WASHER(S) END PLAY (0.005–0.015)

THRUST WASHER

THRU BOLTS (TORQUE 40–55 IN. LBS)

BENDIX ASSEMBLY

LOW MOUNT STARTER

BENDIX DRIVE RETAINING NUT (130–150 IN. LBS)

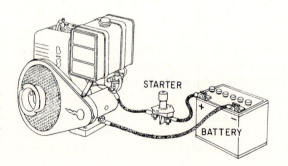

FIGURE 13.2 Belt-driven starter mounted on engine, and connections between the starter and the battery. (*Briggs and Stratton Corporation.*)

STARTER

BATTERY

The engine may have internal trouble such as broken rings or piston, which prevent normal cranking. A following chapter describes engine service.

The fourth condition to consider is the temperature. As the temperature falls, the battery becomes less effective and the engine oil gets thicker. The battery is less effective at low temperatures because it is a chemical device and chemical actions are slowed by low temperatures. Engine oil thickens up with low temperatures, making it harder for the starter to spin the crankshaft. Engine manufacturers point out that with very low temperatures, it may become impossible to start the engine in the normal manner. One manufacturer states, for example, that if the temperature falls below 20° F, you will probably have to use the rope starter to get the engine going.

13.3 Servicing the 12-Volt Starter. Most of these starters are pre-lubricated and require no further lubrication during normal service. About the only trouble that can occur inside the starter results from worn brushes or commutator. You also should consider the possibility that the starter was overloaded from excessively long cranking periods. Recommendations are that the starter should not be used more than about 10 sec at a time. If the engine does not start within this time, the starter should be stopped and allowed to cool off for a minute or so. Continuous cranking will burn out the starter.

When you run into slow cranking, or no cranking at all, and low temperature is not to blame, the first step is to check the battery as we will explain later. Then look at the cables and connections. If cables are frayed and connections are bad, not enough current can get through to produce normal cranking.

On systems using a starter switch or a solenoid, bypass the switch or solenoid by connecting a heavy jumper wire to the two terminals. If the starter then works, you have found the trouble—it is in the switch or solenoid.

If everything outside of the starter looks okay, then check the starter brushes and commutator. Some starters have a cover band which can be removed. On other starters, such as the one shown in Fig. 13.1, you must remove the end cap assembly. This is done by removing the two thru bolts so that the cap can be slipped part way off. If you want to take the cap completely off, you will have to take the insulated brush out of the brush holder. This is done by lifting the spring and gently pulling the brush out.

Now examine the brushes and commutator. If the brushes are worn down, they must be replaced. If the commutator is worn or rough, it should be cleaned with fine sandpaper or a brush-seating stone

(Fig. 13.3). Never use emery cloth to clean the commutator. Some manufacturers supply brush replacement kits. The grounded brush lead is riveted to the end cap, and the rivet must be drilled out so the new brush lead can be riveted into place. The insulated brush is connected to the field lead and must be unsoldered to remove it. Then the new brush lead is soldered to the field lead. Rosin, not acid, soldering flux must be used.

Some manufacturers say that cleaning of the commutator and replacement of brushes are the only services that should be attempted. If there are other troubles in the starter, a new starter should be installed. Complete disassembly of the starter, in other words, is not recommended.

Other manufacturers permit disassembly of the starters in the field and supply detailed instructions and illustrations, such as Figs. 13.4 and 13.5, which show completely disassembled starters. For detailed instructions on how to service them, refer to a book in the McGraw-Hill Automotive Technology Series, "Automotive Electrical Equipment." These are the type of starter found in automobiles and are not commonly used on small-engine equipment.

If the starter is disassembled, then the armature can be put in a lathe so the commutator can be turned down, if the commutator is

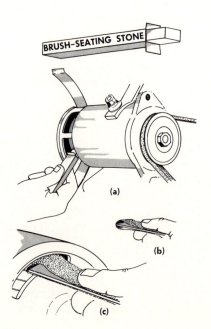

FIGURE 13.3 Cleaning commutator with a brush-seating stone (*a*) or sandpaper (*b* and *c*).

rough or worn and requires this service. See Fig. 13.6. The mica should then be undercut, as shown in Fig. 13.7.

If you are working on a small starter, such as shown in Fig. 13.1, and have removed only the end cap to check the commutator and brushes, then lubricate the bushing and armature shaft before putting the end cap back on. Coat the bushing in the end cap and the end of the armature shaft lightly with SAE 10 oil before installing the end cap. Do not put on too much oil or it will get onto the commutator and brushes and cause trouble.

When putting the end cap back on, you must make sure that the brushes are lifted so they will allow the commutator to pass under them. This can be done by pulling up on the brush springs with needle-nosed pliers. After the cap is partly on, you can let the springs down so the brushes will rest on the commutator. The end cap can then be pushed on into place and the thru bolts installed to complete the assembly.

To inspect the Bendix drive assembly, remove the starter from the engine. If the drive pinion or splined sleeve is damaged, the assembly must be replaced. Do not lubricate the Bendix drive; this can cause it to stick.

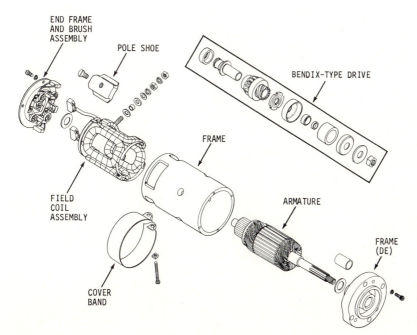

FIGURE 13.4 Disassembled view of an electric starter with a Bendix drive. (*Kohler Company.*)

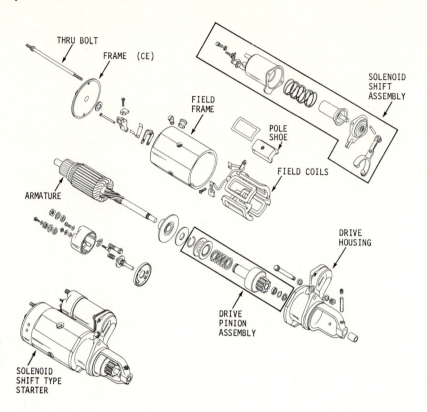

FIGURE 13.5 Disassembled view of an electric starter using an overrunning clutch drive which is actuated by a solenoid. (*Kohler Company*.)

FIGURE 13.6 Turning the armature commutator in a lathe.

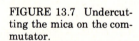

FIGURE 13.7 Undercutting the mica on the commutator.

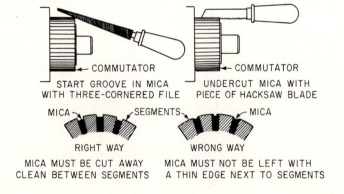

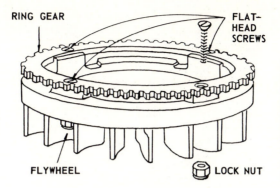

FIGURE 13.8 Method of replacing ring gear in one model of small engine. The old ring gear is removed by drilling out the attaching rivets and then the new ring gear is attached with screws and lock nuts, as shown. (*Briggs and Stratton Corporation.*)

While the starter is off, check the ring gear on the engine flywheel. If the teeth are battered or broken, the ring gear should be replaced. One method of replacing the ring gear is shown in Fig. 13.8. This ring gear is attached with rivets on the original assembly. The rivets must be drilled out as follows: Mark the centers of the rivets with a center punch. Then drill out the rivets with a $\frac{3}{16}$-in. drill. Clean the holes after drilling out the rivets. Then attach the new ring gear with the four screws and lock nuts that are supplied in the ring-gear kit.

When reinstalling the starter on the engine, be sure to use the special mounting bolts and lock washers that you found on the original assembly. These bolts provide the proper alignment of the drive gear to the ring gear (see Fig. 13.9). Incorrect alignment can cause gear clash and damage to the gears.

On belt-drive starters, adjust the belt tension when reinstalling the starter. See Fig. 13.10. The Briggs and Stratton recommendations are to loosen nuts *A* and *B* slightly so the starter can be moved by hand. Then move the starter away from the engine as far as possible. Rock the engine pulley back and forth and at the same time slowly

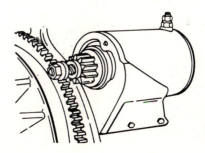

FIGURE 13.9 Relationship of ring gear to gear of Bendix drive. Incorrect alignment will cause poor meshing and rapid tooth wear. (*Kohler Company.*)

slide the starter motor back toward the engine until the starter motor pulley stops being moved by the vee belt. Move the starter another $\frac{1}{16}$ in. toward the engine. Then tighten the nuts *A* and *B*.

13.4 Servicing the 120-Volt Starter. One type of 120-volt starter uses an alternating-current motor that can be connected directly to the 120-volt lighting system in the home, as shown in Fig. 9.8. The other type uses an automotive-type starter which operates on direct current. This starter requires a rectifier to change the alternating current of the house electrical system to the direct current the starter requires to operate. Figure 13.11 is a wiring diagram of the system. Figure 13.12 is a disassembled view of the starter used with the rectifier, shown in Fig. 13.11. The numbers in Fig. 13.12 indicate the order in which the starter parts are to be removed from the assembly. For instance, parts No. 2 and 3, the thru bolts and washers, are removed to permit the drive end cap and assorted other parts to be removed. These parts are all numbered 4 because they all come off together. However, they can be separated in the order of the letters following the number 4. That is, No. 4A, the elastic stop nut, comes off first,

FIGURE 13.11 Wiring diagram of starter which uses rectified house current. (*Tecumseh Products Company.*)

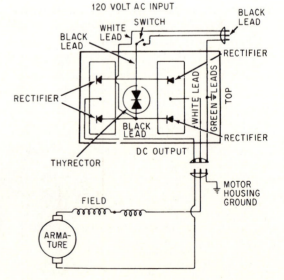

FIGURE 13.10 Method of adjusting drive belt for starter. (*Briggs and Stratton Corporation.*)

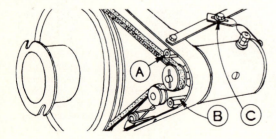

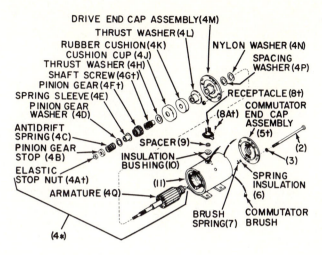

DRIVE END CAP ASSEMBLY(4M)
THRUST WASHER(4L)
RUBBER CUSHION(4K)
CUSHION CUP (4J)
THRUST WASHER (4H)
SHAFT SCREW(4G†)
PINION GEAR(4F†)
SPRING SLEEVE(4E)
PINION GEAR
WASHER (4D)
ANTIDRIFT
SPRING(4C)
PINION GEAR
STOP (4B)
ELASTIC
STOP NUT (4A†)
ARMATURE (4Q)
(4*)

NYLON WASHER (4N)
SPACING
WASHER (4P)
RECEPTACLE (8†)
COMMUTATOR
END CAP
ASSEMBLY
(5†)
(8A†)
(2)
(3)
SPACER (9)
INSULATION
BUSHING(10)
SPRING
INSULATION
(6)
(11)
BRUSH
SPRING(7)
COMMUTATOR
BRUSH

FIGURE 13.12 Disassembled view of starter using 120-volt direct current. (*Tecumseh Products Company.*)

followed by the pinion gear stop No. 4B, the antidrift spring No. 4C, and so on. This illustration makes it relatively simple to disassemble and reassemble the starter. A similar starter is shown disassembled in Fig. 13.13.

13.5 Starter-Generator Service. Generators are used to put back into the battery the current used in cranking, and also to supply cur-

FIGURE 13.13 Disassembled starter used on a small engine. This starter has a rubber-compression-type drive. (*Delco-Remy Division, General Motors Corporation.*)

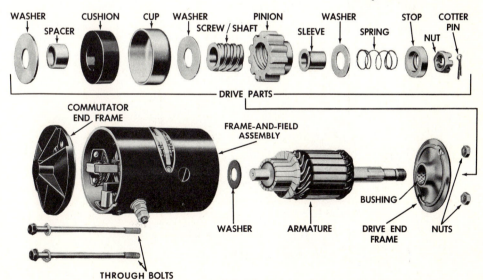

WASHER SPACER CUSHION CUP WASHER SCREW/SHAFT PINION WASHER SLEEVE SPRING STOP NUT COTTER PIN

DRIVE PARTS

COMMUTATOR
END FRAME
FRAME-AND-FIELD
ASSEMBLY
BUSHING
WASHER ARMATURE DRIVE END
FRAME NUTS
THROUGH BOLTS

rent for lights and other electrical equipment that is connected. In some small engines, the generator is a separate unit. But more often small engines will have a single unit that combines the generator and the starter. Inasmuch as the separate direct-current generator is not commonly used on small engines, we will not cover it in detail but will, instead, concentrate on the much more common starter-generator. Actually, servicing procedures are very similar for the two because they are very much alike in construction. Figure 13.14 is a sectional view of a starter-generator, and Fig. 13.15 shows this unit disassembled.

NOTE: Many small engines use a separate alternator to produce the charging current. However, as we have previously noted, in Sec. 9.13, the alternator is usually built into the engine as part of the flywheel assembly. We will cover service to alternators in a later section.

1. Starter-Generator Services Needed. Because the starter-generator is in continuous operation all the time that the engine is running, it should be checked periodically. The tougher the operating conditions, the more frequently the starter-generator should be checked. For instance, frequent starts and stops, excessively long cranking periods due to hard starting, excessively dirty or moist operating conditions, heavy vibration from the engine or surrounding machinery—all these make it necessary for the starter-generator to be checked frequently. In checking, you should look at the brushes, commutator, drive belt, and electrical connections. Following are the recommendations from one manufacturer which are typical.

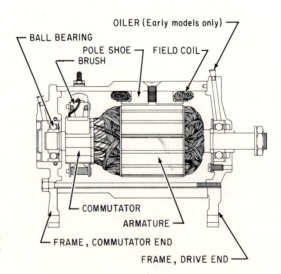

FIGURE 13.14 Sectional view of a starter-generator. (*Kohler Company.*)

2. Brushes. Brushes should be checked about every 200 hr of operation. On the unit shown in Fig. 13.15, this requires removal of the two thru bolts and the commutator end frame. Brushes should be making good contact with the commutator and not be excessively worn. If they are worn to less than one-half of their original length, they should be replaced. Proper brush-spring tension is important and

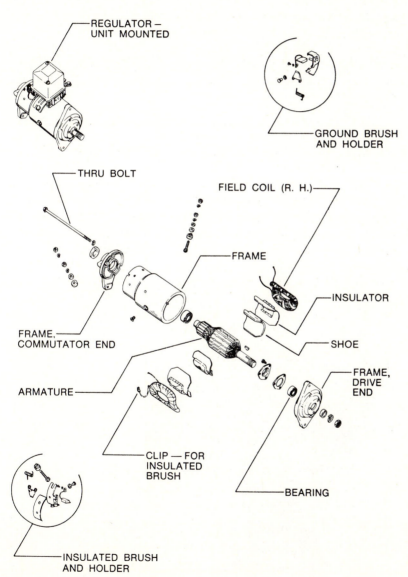

REGULATOR — UNIT MOUNTED

GROUND BRUSH AND HOLDER

THRU BOLT

FIELD COIL (R. H.)

FRAME

INSULATOR

FRAME, COMMUTATOR END

SHOE

ARMATURE

FRAME, DRIVE END

CLIP — FOR INSULATED BRUSH

BEARING

INSULATED BRUSH AND HOLDER

FIGURE 13.15 Disassembled view of a starter-generator. (*Kohler Company.*)

can be measured with a spring scale, as shown in Fig. 13.16. If the tension is not correct, or if the springs appear blued or burned, the springs should be replaced.

3. Commutator. If the commutator is glazed or dirty, it can be cleaned with #00 sandpaper. One way of applying the sandpaper to the commutator is to put the armature in a lathe and, while it is rotating, hold the sandpaper against the commutator. On starter-generators with a cover band or windows in the end frame, it is possible to apply the sandpaper, as shown in Fig. 13.3, while the engine is running and driving the starter-generator. If the commutator is rough, out-of-round, or has high mica, it should be turned down in a lathe and the mica should be undercut, as shown in Figs. 13.6 and 13.7.

> CAUTION. **Never use emery cloth to clean the commutator. Particles of emery can become embedded in the commutator and cause rapid brush wear.**

4. Drive Belt. Make sure that the drive belt is in good condition and is adjusted to the proper tension. Low belt tension will allow the belt to slip so that poor cranking and low generator output will result. Belt slippage will quickly wear out the belt. Belt tension is correct if you can push the belt in $\frac{1}{2}$ in. halfway between pulleys, as shown in

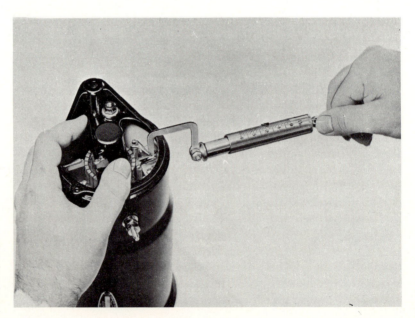

FIGURE 13.16 Measuring spring tension through the opening in the end plate of the generator with a spring scale. (*Delco-Remy Division, General Motors Corporation.*)

Fig. 9.20. Adjust by loosening the bolt that holds the starter-generator tight and swing the unit out to get the correct tension. Then tighten the bolt.

5. Lubrication. Some starter-generators are equipped with sealed ball bearings and require no lubrication. Other units have hinge-cap oilers which should have 8 to 10 drops of light engine oil every 100 hr of operation. Whenever a starter-generator is disassembled, the bearings should be examined for wear. If worn, they should be discarded and new bearings installed. Sealed ball bearings should not be washed in solvent because this would remove the lubricant, and the bearings could not then be relubricated and reused. Unsealed ball bearings, however, should be washed in lubricant, dried, and relubricated, and reused, provided they are in good condition.

6. Starter Tests. If the starter-generator fails to crank the engine properly, check the battery, connections, and wiring. A later section covers battery checking. Make sure the belt is properly tensioned and in good condition so it can carry the cranking effort to the engine. Failure to crank at all, even though the battery and wiring are in good condition, could be due to defects inside the starter-generator or to a defective switch or solenoid. Use a heavy jumper lead to connect around the solenoid or switch. If the starter-generator now cranks the engine, you have found the trouble. If nothing happens, then you should look into the starter-generator for the cause of trouble. Check the commutator and brushes, as previously noted. You will probably have to remove the starter-generator from the engine and disassemble it. Check for tight, worn, or dirty bearings, and a bent armature shaft or loose pole shoes, which could allow the armature to drag. If the armature turns freely, it could be that there is an open circuit in the starter field windings or in the armature. These can be checked with a test lamp and test points.

There is a possibility that something is broken inside the engine which would prevent normal cranking. You can check for this by attempting to turn the flywheel or pulley by hand with the starter-generator loose in its mounting. If you cannot turn the pulley, or if it turns very hard when not on the compression stroke, then the engine must be disassembled for service. A later chapter covers this in detail.

7. Generator Tests. If generator output is zero, first check to make sure the groundstrap from the voltage regulator to frame is not broken or disconnected. Then check the commutator, brushes, and internal connections. Sticking brushes, a dirty or gummy commutator, or poor connections may prevent generator output. If everything appears in good condition in the generator, then it is possible that the cause of trouble is in the regulator and it should be checked, as explained in a following section.

13.6 Current-voltage Regulator Service.   Figure 13.17 shows one model of this regulator with the cover removed so the two units inside can be seen. There are several models of this type of regulator used, as well as some three-unit regulators, such as shown in Fig. 9.25. When you have the job of checking and adjusting these units, you should always refer to the manufacturer's instructions because the procedures required, and the specifications to which the regulator should be set, vary with different models. As a typical example of how you check and adjust a regulator, let us look at the checking procedures for the regulator shown in Fig. 13.17.

1. Regulator Connections.   The connections to the regulator, including the ground connection, must be tight and clean. A loose ground connection, for example, will cause poor regulator operation.

2. Contact Points.   If the regulator contact points become dirty and burned, there will be excessive resistance in the generator field circuit

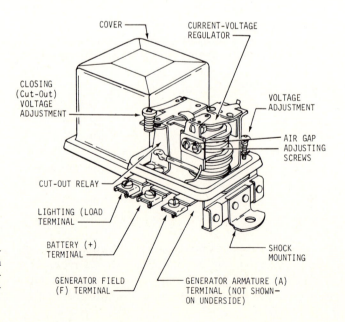

FIGURE 13.17   Current-voltage regulator with cover removed to show adjustments required. (*Kohler Company.*)

COVER

CURRENT-VOLTAGE REGULATOR

CLOSING (Cut-Out) VOLTAGE ADJUSTMENT

VOLTAGE ADJUSTMENT

AIR GAP ADJUSTING SCREWS

CUT-OUT RELAY

LIGHTING (LOAD) TERMINAL

BATTERY (+) TERMINAL

SHOCK MOUNTING

GENERATOR FIELD (F) TERMINAL

GENERATOR ARMATURE (A) TERMINAL (NOT SHOWN ON UNDERSIDE)

with the result that generator output will be low. This could lead to a run-down battery. You can clean the contacts with a fine-cut file. This requires loosening of the screws holding the upper contact assembly in place so it can be swung to one side, or removed. See Fig. 13.18.

> CAUTION. **Never use sandpaper or emery cloth to clean contacts. Particles of sand or emery can become embedded in the contact which will prevent normal regulator action.**

3. Cutout Relay. The cutout relay requires three checks and adjustments; air gap, point opening, and closing voltage. To make the air-gap and point opening checks, disconnect the battery.

a. Air Gap. See Fig. 13.19. Put a finger on the armature directly above the center of the winding and press the armature down until the points just close. Measure the air gap between the armature and the center of the winding core. On the model shown, this should be 0.020 in. Adjust by loosening the two screws attaching the armature hinge to the frame and raise or lower the armature assembly.

b. Point Opening. Release the armature and measure the point opening, as shown in Fig. 13.19. On the model shown, this should be 0.020 in. Adjust by bending the upper armature stop.

c. Closing Voltage. Reconnect the battery, and connect a voltmeter from the generator terminal to ground. Start the engine and

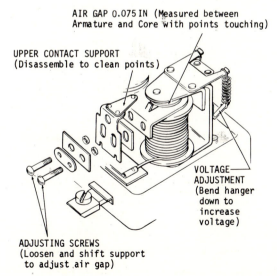

AIR GAP 0.075 IN (Measured between
Armature and Core with points touching)

UPPER CONTACT SUPPORT
(Disassemble to clean points)

VOLTAGE
ADJUSTMENT
(Bend hanger
down to
increase
voltage)

ADJUSTING SCREWS
(Loosen and shift support
to adjust air gap)

FIGURE 13.18 Current-voltage regulator unit, showing adjustments. (*Kohler Company.*)

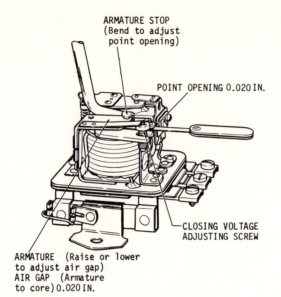

ARMATURE STOP
(Bend to adjust
point opening)

POINT OPENING 0.020 IN.

CLOSING VOLTAGE
ADJUSTING SCREW

ARMATURE (Raise or lower
to adjust air gap)
AIR GAP (Armature
to core) 0.020 IN.

FIGURE 13.19 Cutout relay adjustments. (*Kohler Company.*)

slowly increase its speed, noting the increase in voltage. Watch for the points to close and read the voltage at this instant. The voltage should be adjusted, if it is not correct, by turning the closing-voltage adjusting screw (see Fig. 13.19). Turning it up to increase the tension of the flat spring increases the closing voltage. Recheck the closing voltage after each adjustment by slowing the engine until the points open, and then increasing the speed again.

4. Current-voltage Regulator. This regulator requires two adjustments, air gap and operating voltage.

a. Air Gap. See Fig. 13.18. Adjust by loosening the two adjusting screws and raising or lowering the upper contact point assembly. Adjustment is correct when the specified air gap is attained with the contact points just touching. Tighten the screws and recheck the air gap.

b. Operating Voltage. This regulator operates on both generator voltage and generator output current. The unit must be hot—at operating temperature, before the operating voltage is checked. This requires that the engine be operated for about 20 min so the regulator can warm up. Make the operating voltage check with the regulator cover on. Inasmuch as the voltage varies greatly with the current output, you must measure the operating voltage at a specified current output. For example, one model regulator, with a 3-amp current output, will regulate at 14.4 volts. However, with a 6-amp current output, it will regulate at only 13.2 volts.

Refer to the manufacturer's instructions for the specific model under test so you will know how to connect the meters to make the test. Adjust by bending the lower spring hanger down to increase the voltage setting, or up to lower the voltage setting. Some models have a screw adjustment, similar to that shown in Fig. 13.19 for the cutout relay. Turning the screw in or out changes the voltage setting.

After each adjustment, put the regulator cover back into place, slow the engine until the contact points of the cutout relay open, and bring the engine back up to speed, and then recheck the voltage setting.

13.7 Small-Engine Alternator Service. In automobiles, the alternator is a sizable unit mounted on one side of the engine and driven by the fan belt. Some small engines also use alternators to furnish current for charging the battery and handling external electrical loads, such as lights. These alternators, as we have explained in a previous chapter, are not separately mounted but are built into the engine itself. They are often called flywheel alternators because they use the flywheel as part of the alternator. The flywheel has a series of magnets which whirl past the stationary coils—the stator coils—located around the edge of the flywheel. We have explained how flywheel alternators work in an earlier chapter. Now, we will describe procedures for diagnosing and correcting troubles in small-engine alternators.

NOTE: The automotive-type alternator, which is a separate unit, is serviced in a different manner. For servicing procedures on these, see a book in the McGraw-Hill Automotive Technology Series, "Automotive Electrical Equipment."

Inasmuch as the flywheel alternator has no separate moving parts, it rarely requires service and seldom causes any trouble. We will see what possible troubles might occur with flywheel alternators and how to find out what is causing the trouble. There are certain precautions to observe when working on alternators, as follows:

1. Do not connect the battery backward. If you do this, a high current will flow through the diodes and alternator which would probably burn them up.

2. When recharging the battery from an external source, be sure to disconnect the battery from the rectifier and alternator. This will protect them from damage.

3. When disconnecting the battery, disconnect the grounded terminal first. This avoids damage in case you accidentally ground the insulated terminal with your pliers when loosening the insulated-terminal cable. There is more on battery service on a later page.

4. When checking the system, first note the condition of the leads and connections. Make sure that there is a good ground between the regulator-rectifier and the mounting.

Following are procedures to use to check out a typical system which uses a regulator-rectifier. There are many different arrangements, and it is necessary to look up the specifications and specific procedures to use for any model in the manufacturer's service manual. With the correct information, you will be able to make the checks the correct way. Incorrect checking procedures can lead to damaged equipment.

The system for which the following checks are to be made is shown in Fig. 13.20. There are two basic conditions that require

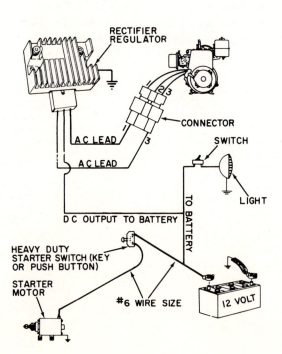

FIGURE 13.20 Wiring circuits for a 7-amp flywheel alternator. (*Tecumseh Products Company.*)

checking—no charging to the battery, and the battery being over-charged.

NO CHARGE TO BATTERY

If no charge is going to the battery and the battery is low (check battery to be sure), proceed as follows to find the cause of trouble. Disconnect the $B+$ terminal wire and connect a direct-current volt-meter from this lead to the case of the regulator-rectifier, as shown in Fig. 13.21. Run the engine near full throttle and read the voltage. If it is above 14 volts, the system is okay and the trouble may be in the ammeter or connections in the circuit.

If the voltage is less than 14 volts but greater than zero, there is probably some defect in the regulator-rectifier. You can try another regulator-rectifier to see if this clears up the trouble, or you can check further, as explained below.

If you get no voltage at all, then the trouble can be in either the stator or the regulator-rectifier, and you will have to isolate the trouble, as explained in a following paragraph.

Check further by reconnecting the battery and checking the volt-age between the two battery terminals, as shown in Fig. 13.22, with the engine operating near full throttle. If the voltage is 13.8 volts or higher, turn on a load such as the lights, to reduce the voltage below 13.6 volts. If the charging rate increases, the system is okay. If the

FIGURE 13.21 Connecting voltmeter leads to check system voltage. (*Tecumseh Products Company.*)

FIGURE 13.22 Checking voltage at the battery. (*Tecumseh Products Company.*)

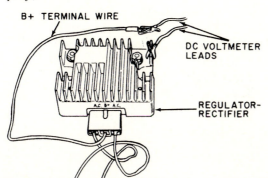

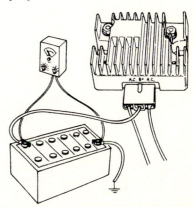

charging rate does not increase, the stator or regulator-rectifier is at fault and a further check must be made.

NOTE: If the system has no ammeter, connect a test ammeter into the circuit when making the above test.

Disconnect the plug from the regulator-rectifier and test the ac voltage (use an alternating-current voltmeter), as shown in Fig. 13.23, with the engine running near full throttle. If the voltage reads less than 20 volts, the stator is defective. If the voltage reads more than 20 volts, the regulator-rectifier is defective.

If tests indicate the regulator-rectifier is at fault, it should be replaced. This is a sealed unit and it cannot be serviced in the field. If tests show the stator is at fault, the trouble probably is due to an open or ground. The coils must be replaced, and this requires partial disassembly of the engine. There is also the possibility that the flywheel magnets have weakened (a rare occurrence), and they can be tested, as shown in Fig. 14.32. See also Sec. 14.7. Procedures vary for the disassembly of different engines, so always refer to the shop manual covering the engine being serviced. There is a general disassembly procedure on small engines in a later chapter in this book.

BATTERY OVERCHARGING

Chances are this is due to a defective regulator-rectifier. You can check with a dc voltmeter connected as shown in Fig. 13.22, with the engine operating near full throttle. If the voltage goes over 14.7 volts, the regulator is not functioning and the regulator-rectifier must be replaced with a new unit. If the voltage remains under 14.7 volts, the

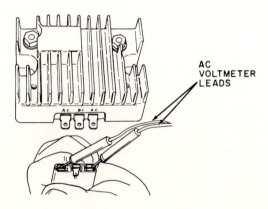

FIGURE 13.23 Checking voltage of stator winding with an ac voltmeter. (*Tecumseh Products Company.*)

AC VOLTMETER LEADS

system is functioning okay. There may be some trouble in the battery, such as a shorted cell, which causes the charging rate to remain high. Battery checking is described on following pages.

13.8 Battery Service.   Battery service can be divided into two parts, regular maintenance to keep the battery in good condition, and checking a battery that has troubles. If you have battery trouble, you want to find out what is causing it, so it can be fixed.

Regular battery maintenance includes periodic checks to determine battery state of charge, adding water, making sure connections are clean and tight, and cleaning off the battery top.

13.9 Battery Testing.   There are several ways to test the battery to find out whether it is in a charged condition or not. The most common is the hydrometer test. Others include the light-load test, the high-discharge test, the instrument or 421 test, and the cadmium-tip test. These other methods require special equipment and we will not go into them here, although a book in the McGraw-Hill Automotive Technology Series, "Automotive Electrical Equipment," covers them in detail. Instead, we will concentrate on the hydrometer test which requires only an inexpensive hydrometer.

As we mentioned on an earlier page, the battery electrolyte contains sulfuric acid. The acid content varies with the battery state of charge. When the battery is charged, the electrolyte contains about 40 percent acid. When it is run down, it has only a few percent. Sulfuric acid weighs about twice as much as water, and this means that the weight, or thickness, of the electrolyte will vary with the state of charge. The hydrometer measures this weight. It contains a float inside the glass tube, as shown in Fig. 13.24. The float will float high in a thick liquid and will sink low in a thin liquid. To use the hydrometer, you remove the battery filler caps, insert the rubber tube into a cell, squeeze the bulb, as shown in Fig. 13.25, release it, and allow electrolyte to be sucked up into the glass tube.

NOTE: If the electrolyte is too low in the cell, add water, as explained in Sec. 13.10, and charge the battery for an hour to mix the water with the electrolyte in the battery. Then check the gravity with the hydrometer.

CAUTION.   The sulfuric acid in the electrolyte is extremely corrosive and will attack almost everything it comes into contact with. It will eat holes in clothing, and will cause serious skin burns. It can put out your eyes if it gets into them. So, handle the electrolyte with extreme care! If electrolyte is spilled, flush it away with water. Also, baking soda may be thrown on the spilled electrolyte to neutralize it and render it harmless. Then the neutralized electrolyte should be washed away.

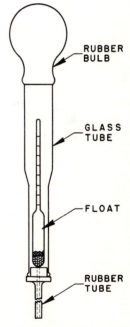

FIGURE 13.24 Battery hydrometer for measuring specific gravity of battery electrolyte, and thus the state of charge, of batteries.

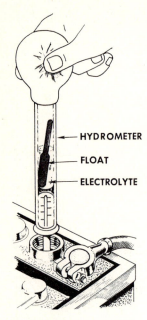

If electrolyte gets on the skin, wash it away with water, followed by soap and water. If electrolyte gets in the eyes, wash the eyes out with water, repeatedly. Then see an eye doctor as soon as possible.

When sucking electrolyte into the hydrometer glass tube, do not suck in too much. Take in just enough to allow the float to float freely. Make sure the float does not stick to the side of the glass tube. Take your reading at eye level, as shown in Fig. 13.26, noting the marking on the float stem at the point where it is level with the electrolyte. The more sulfuric acid there is in the electrolyte, the thicker the electrolyte, and the higher the battery state of charge. The thicker the electrolyte, the higher the float will rise. The float stem is marked off to indicate the thickness of the electrolyte. It is indicated in terms of specific gravity and the markings vary from about 1.100 which indicates a fully discharged battery to 1.300 which indicates a fully charged battery. Actually, specific gravities are not exact measurements of the state of charge. The table below gives approximate measurements.

FIGURE 13.25 Using a hydrometer to check the specific gravity of battery electrolyte.

Approximate Gravity	State of Charge
1.260–1.290	Fully charged
1.230–1.260	About three-fourths charged
1.200–1.230	About half charged
1.170–1.200	About one-fourth charged
1.140–1.170	About run down
1.110–1.140	Discharged

If the temperature of the electrolyte is unusually hot or unusually cold, you should take this into account in interpreting the specific gravity reading. The reason is that when the electrolyte gets cold it thickens up. When it gets hot, it thins out. This changes the specific gravity. To compensate for this, you have to subtract gravity points from your reading if the temperature is low, or add gravity points if the temperature is high. For every 10° the temperature changes, the specific gravity changes about four thousandths (0.004). The reference point is 80°.

For example, suppose you took a reading of the electrolyte which is at 0° and it measured 1.230. To correct this reading for temperature, you would have to subtract 0.032 (8 times 0.004) to get a corrected reading of 1.198. So what you might think is a half-charged battery turns out to be, when you correct the gravity reading, only about a quarter charged.

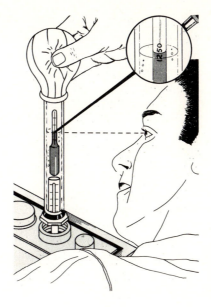

FIGURE 13.26 Taking the specific gravity reading on the stem of the float.

Normally, you don't have to worry about making this correction, provided the electrolyte temperature is not too far from the 80° standard.

One thing to consider, however, in freezing weather, is the possibility of the battery freezing. A fully charged battery will not freeze until the temperature drops down to about −95°, which is probably colder than it will ever get where you live. However, a run-down battery will freeze at around +18°. So it is very important to keep the battery charged during cold weather. The table below shows you various gravities and their freezing temperatures.

Gravity	Freezing Temperature, °F
1.100	18
1.160	1
1.220	−17
1.230	−31
1.260	−75
1.300	−95

Don't make the mistake that some have made, however, by bringing the battery inside out of the cold and allowing it to sit in a warm place for a long period without recharging it. A warm battery will slowly self-discharge. For example, a battery kept at 100° for 3 months will almost completely self-discharge. However, if it is kept at 60° F it will lose only about a fourth of its charge in 3 months. If you do store the battery in a warm place, you should see to it that the battery gets a charge every month or so.

After checking the specific gravity of a battery cell, put the electrolyte back into the same cell from which you took it. Check all battery cells, noting their specific gravities. If all are about the same, and they all measure around 1.230 to 1.300, then the battery can be assumed to be in a charged condition.

NOTE: In referring to specific gravities, you ignore the decimal point and say, for instance, twelve twenty-five (for 1.225) gravity, or eleven fifty (1.150) gravity.

If one cell is low with the others having considerably higher gravity, chances are the low cell is shorted. This means that a new battery is required, or soon will be.

If all cells measure low, then the battery should be given a charge from an external charger. It is not good for a battery to be allowed to sit around in a discharged condition. Sulfate that forms in the plates during discharge tends to crystalize after a time, and this is very hard to break down. That is, the battery is hard to recharge after it has been sitting around in a discharged condition for a time.

13.10 Adding Water. If the electrolyte is low in the battery cells, water should be added. It is best to use distilled water, but any water that is clean and free of impurities will be okay. Remember, however, that the more chemicals, such as lime or iron, there are in the water, the worse it is for the battery. Water heavily loaded with iron, for example, will greatly shorten battery life.

NOTE: It is important to keep the electrolyte level above the battery plates. Plate area exposed to air will be damaged so that battery capacity will be lost. Continual battery operation with low electrolyte can ruin the battery.

When adding water, avoid overfilling the cells. Many batteries have nonoverfill devices, such as shown in Fig. 13.27. When water is added, so the electrolyte level rises to contact the slotted lower end of the vent well, the electrolyte surface will distort where it touches the vent well. This shows that the electrolyte level is correct and no more water should be added.

13.11 Cleaning Battery. Moisture and dirt collecting on the top of the battery can form a discharge path between the battery terminals

ELECTROLYTE
LEVEL LOW

ELECTROLYTE
LEVEL NORMAL

SURFACE OF ELECTROLYTE
BELOW SPLIT RING

FILL TO SPLIT
RING

FIGURE 13.27 Nonoverfill battery device. The appearance of the electrolyte and split ring when the electrolyte level is too low and when it is correct. (*Delco-Remy Division, General Motors Corporation.*)

so that the battery will gradually run down. For this reason, the top of the battery should be cleaned periodically. To do the job, tighten the filler plugs, sprinkle or brush baking soda solution on the battery top (to neutralize any acid), wait until the foaming stops, and then flush off the battery top with clean water.

13.12 Cleaning Terminals and Cable Clamps. Accumulation of corrosion around the battery terminals and cable clamps is more or less normal. If you don't do anything about it, the corrosion will ultimately cause poor connections to develop which will result in a rundown battery. To clean the clamps and terminals, first remove the clamps from the terminals.

On the type of clamp using a nut and bolt, loosen the nut about $\frac{3}{8}$ in. Never use a bar or screwdriver to pry the clamps loose; this can break plate connections inside the battery. For a similar reason, do not use ordinary pliers or an open-end wrench to loosen the cable-clamp nuts. The jaws of the pliers or wrench could swing around and break the cell cover. Instead, use the special cable pliers, as shown in Fig. 13.28, which have narrow jaws.

If the clamp sticks to the battery terminal, use a clamp puller, as shown in Fig. 13.29, to pry the clamp loose. This does the job without putting any pressure on internal battery parts.

To detach the spring-ring type of cable clamp, squeeze the ends of the prongs together with a pair of pliers, as shown in Fig. 13.30. This expands the clamp so it can be lifted off.

Use special wire brushes or steel wool to clean the clamps and terminal posts to bright metal. Figure 13.31 shows how special wire brushes are used to clean the terminals and clamps. Be sure to flush off all traces of filings with clean water.

If the cables have worn insulation, frayed wires, or poor connections, they should be replaced. When reinstalling the cables on the battery, use a special clamp spreader, if necessary, as shown in Fig. 13.32, in order to get the clamp down into place correctly, as shown in

FIGURE 13.28 Using battery cable pliers to loosen nut-and-bolt type of battery cable. (*United Delco Division of General Motors Corporation.*)

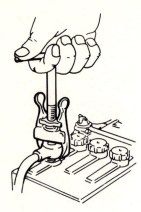

FIGURE 13.29 Using a special clamp puller to pull cable from battery terminal. (*United Delco Division of General Motors Corporation.*)

Fig. 13.33. Never hammer on the clamp to get it down into place. This can break internal connections and ruin the battery. To tighten the nut on the nut-and-bolt clamp, use the special pliers, as shown in Fig. 13.28.

A coating of petroleum jelly or anticorrosion paste on the clamps and terminal posts will retard corrosion.

13.13 Battery Trouble Diagnosis. A battery is built to last a long time. How long it will live depends on the type of service it is in and the kind of care it is given. A battery that is subjected to frequent discharging and recharging will fail sooner than a battery that is kept in a charged condition most of the time. But a battery that is overcharged, or does not get water when it needs it, will have a short life. The two most common complaints, aside from short battery life, are a run-down battery or a battery that requires frequent watering.

1. Battery Uses Excessive Water. When a battery requires frequent addition of water in order to keep the electrolyte above the plates, the chances are it is being overcharged. When a battery continues to receive a relatively high charging rate after it has reached a fully charged condition, it begins to lose water rapidly. The high charge

FIGURE 13.31 Using special wire brushes to clean battery terminal posts and cable clamps. (*Buick Motor Division of General Motors Corporation.*)

FIGURE 13.30 Using pliers to loosen a spring-type of cable clamp from a battery terminal. (*United Delco Division of General Motors Corporation.*)

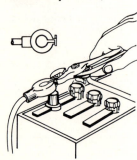

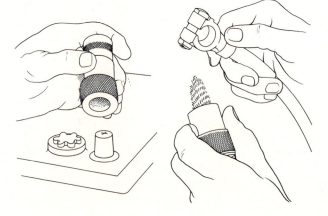

can damage the battery plates and, in some instances, will cause them to swell so much that they will push the positive sides of the cell covers up. The remedy is to reduce the charging rate. If the system has an adjustable regulator, the regulator should be readjusted to a lower voltage, as explained in the manufacturer's shop manuals. On systems such as shown in Fig. 9.31 which have a means of cutting down on alternator output, the switch should be set for low output until the battery gravity starts to fall, showing it needs the higher rate.

A battery that requires water in only one or two cells could have a cracked case which is leaking. You could detect this easily, however, because the battery holder would be wet all the time and would soon show signs of corrosion.

2. Battery Run Down. A run-down battery can be due to external causes such as an inadequate generator or alternator output, or internal causes such as internal shorts or old age. Let us consider various possible causes.

If the generator or alternator does not put out enough current, the battery will run down. The cause could be defects within the generator or alternator, or the regulator may not be adjusted or is not functioning properly. Poor connections in the charging circuit might be preventing a normal charge from reaching the battery. Also, more electrical load may be connected than the generator or alternator can handle, and this puts an extra demand on the battery. If a load, such as lights, is left on when the engine is not running, the battery has to carry the whole load and will soon run down.

If the battery is kept in a hot place, it will self-discharge rapidly when it is not being charged. You should also consider that a dirty battery top may allow discharge current to flow between the battery terminals, and this can run down the battery after a time. Finally, consider that the battery may be so old that it is nearly worn out. That is, it can no longer take and hold a charge in a normal manner. The remedy here is to get a new battery.

FIGURE 13.32 Using a cable-clamp spreader to spread the clamp and thus assure good seating of the clamp on the battery terminal post. (*United Delco Division of General Motors Corporation.*)

FIGURE 13.33 Wrong and right way to put a cable clamp on the terminal post. (*United Delco Division of General Motors Corporation.*)

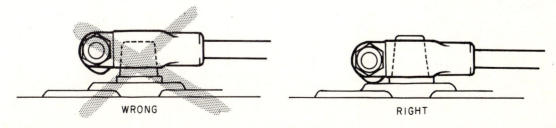

WRONG   RIGHT

CHECKUP

You have now covered the servicing procedures for the various kinds of electric starters used on small engines. As you have learned, there is quite a variety of electric starters, and each requires its own service procedure. Fortunately, many of these procedures are very similar, so that the fundamentals of electric-starter service are relatively easy to learn. The following questions will not only give you a chance to check up on how well you understand and remember these fundamentals, but also will help you to remember them better. The act of writing down the answers to the questions will fix the facts more firmly in your mind.

NOTE: Write down your answers in your notebook. Then later you will find your notebook filled with valuable information which you can refer to quickly.

Completing the Sentences: Test 13.   The sentences below are not complete. After each sentence there are several words or phrases, only one of which will correctly complete the sentence. Write each sentence in your notebook, selecting the proper word or phrase to complete it correctly.

1. If the 12-volt starter cranks the engine slowly or not at all, the cause could be a run-down battery, defective starter, engine trouble, or (*a*) fuel-system trouble; (*b*) ignition-system trouble; (*c*) low temperature.
2. The major wearing parts in the 12-volt starter are the (*a*) bearings and commutator; (*b*) brushes and commutator; (*c*) armature and fields.
3. Two services that can be performed on 12-volt starter armatures are (*a*) rewinding and resoldering; (*b*) replacing commutator and undercutting mica; (*c*) turning commutator and undercutting mica.
4. Starter-generators do two jobs, they start the engine and (*a*) recharge the battery; (*b*) operate the voltage regulator; (*c*) close the cutout relay.
5. The three checks to be made on the cutout relay are air gap, (*a*) armature gap, and point opening; (*b*) point opening, and point closing; (*c*) point opening, and closing voltage.
6. The two checks to be made on the voltage regulator are (*a*) air gap and operating voltage; (*b*) point opening and closing voltage; (*c*) operating voltage and amperage.

7. When a battery is not to be used for a period of time, it should be fully charged and stored in (*a*) a warm place; (*b*) a cool place; (*c*) a freezer.
8. About the only troubles that can occur with a flywheel alternator include loss of flywheel magnetism, defective stator coils, and (*a*) defective cutout relay; (*b*) wrong voltage setting; (*c*) defective diodes.
9. The battery hydrometer measures (*a*) battery voltage; (*b*) battery age; (*c*) specific gravity.
10. If all cells of a battery require frequent additions of water to keep the electrolyte level above the plates, chances are the (*a*) battery is being overcharged; (*b*) case is leaking; (*c*) battery is undercharged.

Written Checkup

In the following, you are asked to write down, in your notebook, the answers to the questions asked or to define certain terms. Writing the answers down will help you to remember them.

1. Describe the procedure of disassembling a 12-volt starter, checking and servicing the parts, and reassembling it.
2. Describe the procedure of servicing a 120-volt starter, referring to to the manufacturer's service manual as necessary.
3. Describe the procedure of servicing a starter-generator.
4. Explain how to check and adjust a current-voltage regulator.
5. Describe the checking and servicing of a flywheel alternator.
6. Explain how to check battery specific gravity and what various gravity readings mean.
7. Explain how to remove and replace battery cables.
8. Explain how to clean the battery top.

Ignition System Service

14

We described in a previous chapter the two basic types of ignition systems used in small engines, magneto ignition and battery ignition. Magneto ignition systems can be further classified as flywheel magnetos and external magnetos. There are also solid-state ignition systems which use a flywheel magneto to supply the pulses of electrical current. All these were described in Chap. 7. Now, let us look at the trouble-diagnosis and servicing procedures for these different types of ignition systems.

14.1 Basic Magneto Maintenance. Without a good spark, delivered to the spark plug at the correct instant, the engine will perform poorly or not at all. Producing a good spark, and delivering it to the spark plug at the correct instant, depends on several factors. The contact points must be in good condition and properly adjusted. The magneto stator coils, condenser, wiring (particularly the high-tension lead to the spark plug), and connections must also be in good condition. And the spark plug must be clean and properly adjusted. Now let us see how to check these various system components, and make corrections or adjustments as necessary to get top performance from the engine.

14.2 Magneto Trouble Causes. In Chap. 10, we described procedures for tracking down the causes of such troubles as failure to start, lack of power, engine surging, engine losing power as it runs, and irregular firing. You will recall how we explained the spark test — seeing whether or not a spark occurs at the spark plug when the engine is cranked. A refinement of this test utilizes a spark tester, as shown in Fig. 14.1. To use this tester, remove the spark plug and connect the high-tension lead from the magneto to the tester, as shown. Ground the electrode of the tester that gives the 0.166 in. gap, as shown, and spin the flywheel. If a good spark jumps the gap, you can assume that the ignition system is functioning normally.

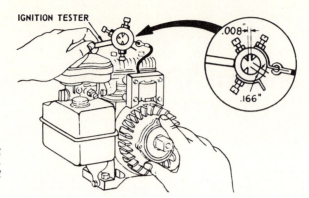

FIGURE 14.1 Checking spark with special spark-gap tester. (*Briggs and Stratton Corporation.*)

If the engine runs but misses, you can check out the ignition system by inserting the tester, shown in Fig. 14.1, between the clip on the high-tension lead and the spark plug. Run the engine and watch carefully for the spark at the tester gap. If the spark is not regular and steady, you can assume that there is trouble in the ignition system; the points, wiring, and stator coils should be checked out.

If you do not get a spark on the spark test, then there are several things to check:

1. Wiring for bad insulation or poor connections.
2. Grounded ignition switch which prevents opening of the primary circuit so that primary current is not interrupted when the contact points open.
3. Shorted condenser which would have the same effect.
4. Magneto coil shorted, open, or grounded so it cannot produce high voltage.
5. Loss of magnetism of permanent magnets on flywheel or in external magneto rotor.

If the engine backfires, or kicks back, when starting, the breaker-point gap may be too wide. This causes the spark to occur too early before the end of the compression stroke so the pressure builds up enough to try to kick the piston back down before it can reach TDC. An early spark will also cause spark knock when the engine is running. The remedy is to reset the timing. On some magnetos, the only adjustment is to change the breaker-point opening. On others, you can also move the breaker plate to change the timing. We will get into that later.

14.3 Removing Flywheel to Check Internal Magneto Points. To check the breaker points of a flywheel magneto, you have to remove the flywheel on many models. On other models, the breaker points are mounted in an external breaker box, as shown in Fig. 14.2. Let us look at the internally mounted type first, which requires removal of the flywheel. We will look at the external type in a later section.

> CAUTION. **Always disconnect the spark plug wire from the spark plug when working on the engine to prevent accidental starting.**

First, if the engine has a rope rewind or a windup starter, remove it as already explained in the chapter on servicing mechanical starters (Chap. 12). If the engine has an electric starter-generator, the flywheel will have a stub shaft, as shown in Fig. 14.3, on which the drive pulley mounts. This stub shaft must be removed. See Fig. 14.3. Then, the flywheel shroud must be removed so you can get at the flywheel.

Next, remove the flywheel attaching nut. Note that some of these nuts have right-hand threads and some have left-hand threads. You can usually determine which is which by considering the normal rotation of the engine. If the flywheel turns clockwise while you are crank-

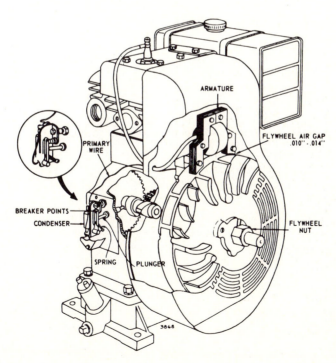

FIGURE 14.2 Engine with magneto using an external box to hold breaker points. Engine has been partly cut away so the camshaft can be seen. The plunger, or push rod, operates the points. (*Briggs and Stratton Corporation.*)

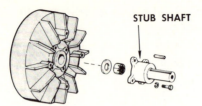

FIGURE 14.3 Stub shaft on which the pulley to drive the starter-generator mounts.

ing it, then the nut has a right-hand thread. Cranking the engine tends to tighten the nut. If the nut had a left-hand thread, then turning the flywheel would tend to loosen the nut.

You will need not only a wrench of the correct size to fit the nut, but also some means of holding the flywheel. Figure 14.4 shows one holder in use while the nut is being loosened. Another type is shown in Fig. 14.5. You can also use a wooden block, as shown in Fig. 14.6, if the engine is mounted on a solid surface so it will not move around when you apply pressure to the wrench.

On some engines, it is possible to loosen the flywheel nut by simply rapping the wrench sharply with a soft hammer (plastic or brass headed) as shown in Fig. 14.7.

Next, you remove the flywheel from the crankshaft. There are several ways of doing this. If the shaft is tapered, you can thread a

FIGURE 14.4 Removing flywheel attaching nut while holding the flywheel with a special holder. (*Tecumseh Products Company.*)

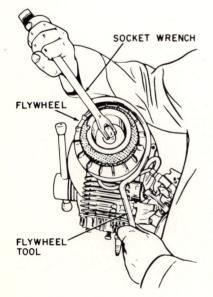

FIGURE 14.5 Using a special tool to hold the flywheel while the starter clutch is being loosened with a special tool. (*Briggs and Stratton Corporation.*)

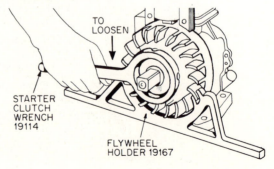

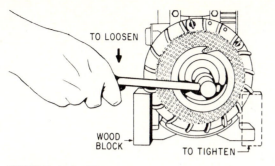

FIGURE 14.6 Nuts on larger flywheels can be loosened by holding the flywheel with a block of wood, as shown. (*Briggs and Stratton Corporation.*)

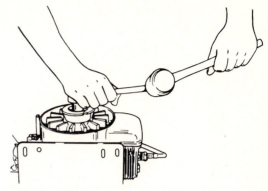

FIGURE 14.7 On some engines it is possible to loosen the flywheel nut by striking the wrench handle with a soft hammer. (*Tecumseh Products Company.*)

nut down on the threads, turning it down so the threads on the end of the shaft are almost exposed. Then rap the nut with a soft hammer, as shown in Fig. 14.8. The nut takes the blow and protects the threads on the shaft. Some manufacturers supply a special puller which serves the same purpose as the nut but provides better protection for the

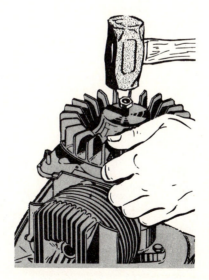

FIGURE 14.8 Showing one way to loosen the flywheel from a tapered shaft.

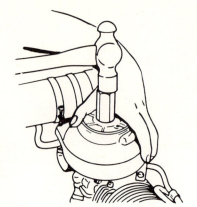

FIGURE 14.9   Using a special flywheel knockout puller to loosen flywheel from shaft. (*Tecumseh Products Company.*)

threads. Figure 14.9 shows this puller in use. It is turned down on the threads until it is about $\frac{1}{16}$ in. from the flywheel. Then it is given a sharp rap with a soft hammer. As a rule, only one or two raps with the hammer on the nut or puller will loosen the flywheel. You must be careful, however, about too much hammering because this could weaken the permanent magnets and also might damage the crankshaft bearings.

If the shaft is not tapered then you will need to use a different sort of flywheel puller—one that has threads which gradually pull the flywheel off the shaft. Figure 14.10 shows one type. Attach the puller with screws, as shown. Then turn the screw handle down. The screw rests on the end of the shaft, and as the screw is turned down in the puller, the pulley is pulled loose from the shaft. There are other kinds of pullers; Fig. 14.11 shows another type, for example.

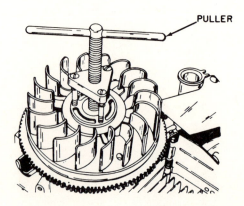

FIGURE 14.10   Using a screw-thread puller to remove flywheel from shaft. (*Tecumseh Products Company.*)

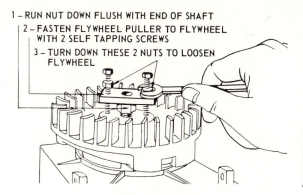

1 – RUN NUT DOWN FLUSH WITH END OF SHAFT
2 – FASTEN FLYWHEEL PULLER TO FLYWHEEL
   WITH 2 SELF TAPPING SCREWS
3 – TURN DOWN THESE 2 NUTS TO LOOSEN
   FLYWHEEL

FIGURE 14.11 Using a screw-type puller to remove flywheel from shaft. (*Tecumseh Products Company.*)

CAUTION. **Do not drop the flywheel or handle it roughly. You can knock most of the magnetism out of the permanent magnets with rough treatment. Also, heat can take the magnetism out of magnets, but normally you would not have to worry about this.**

14.4 Dust Cover. On many engines with internal points, you will find that they are protected by a breaker-point cover, as shown in Fig. 14.12. Remove this cover carefully because if it is bent during removal it would not seal on reassembly and a new cover would be required. Note that the point where the leads come out from under the cover is sealed. This seal is made with Permatex or similar sealer, when the cover is replaced. A poor seal either around the rim of the cover or at the leads will allow dirt to get into the breaker-point compartment with the result that the points will burn and require frequent replacement.

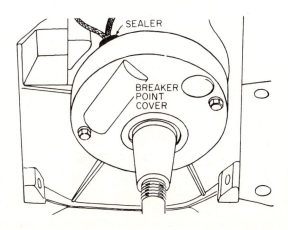

SEALER

BREAKER
POINT
COVER

FIGURE 14.12 Breaker-point cover, also called the dust cover. (*Briggs and Stratton Corporation.*)

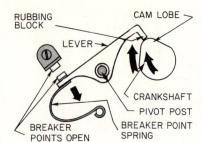

FIGURE 14.13 Arrangement with breaker points being opened by a cam working directly on the breaker lever.

14.5 Breaker-point Service. The breaker points have a stationary point and a movable point on a breaker arm. The design and arrangement of the points differ with different engines. On some engines, the stationary point is on the end of the condenser. On others, it is mounted on a bracket. The breaker arm, on which the movable point is mounted, is pivoted so it can move the point up against or away from the stationary point. There are various arrangements to move the breaker arm. Figures 14.13 to 14.15 show three arrangements. Regardless of the method, when the high point, or lobe, of the cam comes around under the rubbing block (Fig. 14.13), the push rod or plunger (Fig. 14.14), or

FIGURE 14.15 With this arrangement, the cam operates a trip lever which actuates the lever on which one contact is mounted.

FIGURE 14.14 With this arrangement, the cam works a push rod which actuates the lever to open the points.

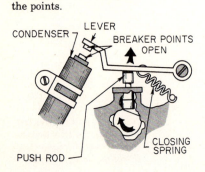

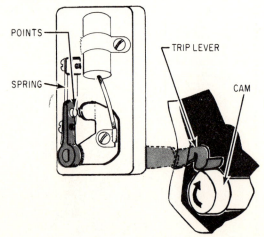

the trip lever (Fig. 14.15), the breaker arm is moved and the contact points separate. Figure 14.2 shows the push rod arrangement for an engine with the breaker points mounted in an external breaker box. The cam is on the crankshaft on two-cycle engines. On four-cycle engines, the cam is on the camshaft.

1. Cleaning Points. Once you have the breaker points exposed by removal of the flywheel and the dust cover, where present, examine them for oxidation or pits. If they are only slightly burned or pitted, they can be cleaned with a special ignition file, as shown in Fig. 14.16. It is not necessary to file the points until they are smooth. Just remove the worst of the high spots. Blow out all dust after cleaning the contacts. Pull a strip of clean bond paper between the points (with points closed) to remove the last traces of filings.

FIGURE 14.16 Breaker points can be cleaned with a contact file.

> CAUTION. **Never use emery cloth or sandpaper to clean the points. Particles of emery or sand will embed in the points and cause erratic operation and possibly point burning.**

2. Installing New Points. If points are badly burned, worn, or pitted, they should be replaced. Badly burned points could be due to a defective condenser, contact points out of adjustment, or oil on the contact surfaces. Check for these conditions before replacing the points. Various methods of attaching the stationary point and breaker arm are used. On the type shown in Fig. 14.17, the point assembly is removed by removing the condenser wire from the breaker-point clip and then loosening the adjusting lock screw so the assembly can be slipped off. On the type shown in Fig. 14.18, the breaker arm is removed by loosening the screw holding the post in position. The stationary contact is on the condenser and is removed along with the condenser by loosening the condenser-clamp screw.

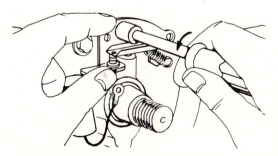

FIGURE 14.17 On this design, the breaker-point assembly is removed by loosening the adjustment lock screw. (*Briggs and Stratton Corporation.*)

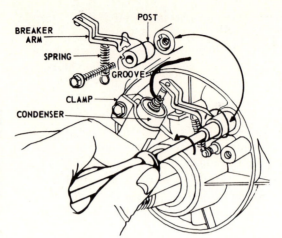

FIGURE 14.18 On this design, the breaker-point assembly is removed by loosening the screw holding the post. The condenser, with stationary point, is removed separately by loosening the condenser-clamp screw. (*Briggs and Stratton Corporation.*)

On engines which use a push rod, or plunger, to operate the breaker arm, the engine manufacturer recommends a check of the plunger and plunger hole whenever the contact points are removed. For example, Fig. 14.19 shows a plug gauge being used to check the hole for wear. If the hole is enlarged, it will allow oil to enter the breaker-point compartment where it will get on the points and cause them to burn and give trouble. If the hole is worn, it must be reamed out and a special bushing installed. The procedure for doing this is shown in Fig. 14.20. The plunger should be checked for wear, and if it is worn down until it is too short, it should be replaced with a new plunger.

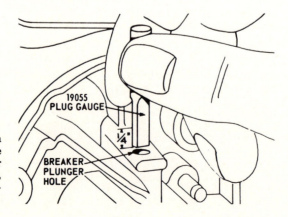

FIGURE 14.19 Using a plug gauge to check the hole in which the breaker push rod, or plunger, works. (*Briggs and Stratton Corporation.*)

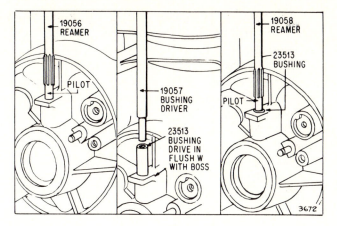

FIGURE 14.20 If the plunger hole is worn, it must be reamed out and a bushing installed, as shown. Then the bushing is reamed out to the proper size. (*Briggs and Stratton Corporation.*)

On the type of breaker arm that uses a rubbing block (Fig. 14.13), check for rubbing block wear. As the rubbing block wears, it increases the breaker-point opening and can change the timing.

Check the breaker cam for wear. Some cams on two-cycle engines are integral with the crankshaft and if they are badly worn, the crankshaft must be replaced. On other engines, the cam is a separate collar locked to the crankshaft by a key. On these, the cam can be replaced if it is worn.

Check for a leaky crankshaft seal because this would allow oil to get on the breaker points so that they would burn rapidly. A leaky seal should, of course, be replaced.

After removing the old points, install the new ones, carefully noting the proper relationship, as shown in Figs. 14.13 to 14.18. Then, check the point opening and adjust it as necessary. Finally, ignition timing should be checked.

3. Adjusting Points. To adjust the breaker points, first make sure they are properly aligned. Figure 14.21 shows the right and wrong ways to align points. As a rule, you will find that the points are properly aligned and no adjustment is required. However, if you do find misalignment of a set of new points, you can adjust them by slightly bending the bracket supporting the stationary point.

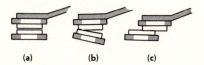

(a) (b) (c)

FIGURE 14.21 (*a*) Correct breaker-point alignment and (*b*) and (*c*) incorrect alignment.

The point opening is a critical adjustment because if it is excessive, the ignition can be too advanced, and this can cause engine backfiring on starting, as well as spark knock when the engine is running.

To adjust the point opening, turn the crankshaft in the direction of normal rotation until the cam opens the points to the widest position. Then use a feeler gauge of the proper thickness to measure the gap between the points. Make sure the feeler gauge is clean so you don't get oil or dirt on the points. The thickness of the feeler gauge selected to make the measurement varies with different engines. Always check the specifications for the engine being checked.

Adjustments are made in different ways, according to the method of point attachment. On some magnetos, the adjustment is made as shown in Fig. 14.22. The lock screw holding the bracket on which the stationary point is mounted is loosened slightly. Then a screwdriver is inserted into the slot, as shown, and twisted to move the stationary point the correct amount to get the proper point opening. Then, the lock screw should be tightened.

Another arrangement is shown in Fig. 14.23. With this arrangement, the stationary point is mounted on the end of the condenser. To make the adjustment, the condenser-clamp screw is loosened slightly, and then the condenser is shifted one way or the other with the screwdriver to get the proper point opening.

In the arrangement shown in Fig. 14.24, the stationary point is mounted on a bracket, and the lock screw must be loosened to shift the bracket as necessary to get the proper point opening. Then, the lock screw is tightened.

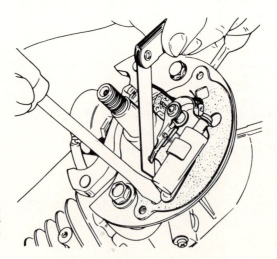

FIGURE 14.22 Typical magneto breaker-point adjustment. (*Tecumseh Products Company.*)

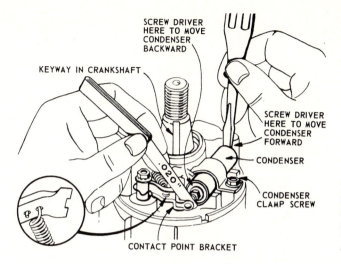

FIGURE 14.23 With this arrangement, the point opening is adjusted by shifting the condenser one way or the other. (*Briggs and Stratton Corporation.*)

14.6 Timing Ignition. To time the ignition means to make an adjustment that will cause the spark to occur at exactly the right moment before the piston reaches TDC on compression. This starts the ignition process at the correct moment so that maximum power will be realized on the power stroke. If the timing is early, the engine will backfire on starting and knock when running. If the timing is late, the power stroke will be weak because ignition will not start until after the piston has begun to move down on the power stroke.

Various methods of timing the ignition are used. Ignition can be timed either with the engine not running (static timing) or with the engine running.

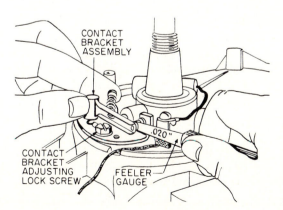

FIGURE 14.24 In this magneto, the point opening is adjusted by loosening the lock screw and shifting the bracket on which the stationary point is mounted. (*Briggs and Stratton Corporation.*)

The sequence, shown in Fig. 14.25, shows the static ignition-timing procedure for one line of small two-cycle engines. After installing the points (1), align them using the special tool shown, to bend the stationary point support (2). Then measure the point opening and adjust it as required (3). Next, clean the points with lint-free paper (4) and use a timing tool or rule to locate the TDC position of the piston, as shown at 5 and 6. Back off the piston by turning the crankshaft backward (7). Find the timing dimensions in the manufacturer's specifications and adjust the tool to that dimension. Then tighten the thumb screw to lock the dimension, as shown in 8 and 9. Next, slowly rotate the crankshaft in the normal running direction until the piston touches the bottom of the tool (10). Then install a 12-volt test light (see Fig. 14.26) connected across the contact points (11). With the stator hold-down screws loosened slightly, shift the stator, as shown in 12 until the points just open. This is indicated by the light going out. Tighten the hold-down screws. As an alternative, you can use a strip of thin cellophane between the points, and this will fall out as the points separate. After completing the timing, install lead, cover, flywheel, and lower housing.

Figure 14.27 shows how to time a running engine by using a timing light. The timing light is connected to the high-tension lead between the magneto and the spark plug. It picks up a signal each time the plug fires and produces a momentary flash of light. During the timing operation, the light is directed to a hole in the engine case through which the flywheel can be seen. The flywheel has a mark on it that should align with a mark on the case when the timing is correct. If the marks do not align, then adjustment is made by loosening the point-opening adjusting screw, and shifting the breaker plate with a screwdriver, as shown to the right in Fig. 14.27. When the marks align, tighten the breaker-plate screw. Note that the ignition timing is made with the engine running at a specified speed. With the engine shown, the specified speed is 1,200 to 1,800 rpm (revolutions per minute).

Some engines have timing alignment marks on the armature mounting bracket and flywheel, as shown in Fig. 14.28. To time the ignition on these models, remove the flywheel, set the points to the proper opening, put the flywheel back in place, and run the nut up only finger tight. Then rotate the flywheel in the running direction until the points are just opening. Next, take off the flywheel, being very careful not to turn the crankshaft at all. Note the positions of the arrows. If they do not align, slightly loosen the mounting screws holding the armature bracket to the engine cylinder. Slip the flywheel back on the crankshaft, using the key to get correct alignment. Install

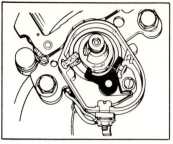

1. INSTALL POINTS

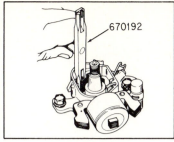

670192

2. ALIGN POINTS

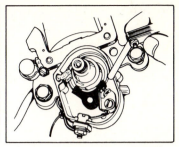

3. POINT OPENING ADJUSTMENT

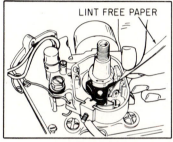

LINT FREE PAPER

4. CLEAN POINTS

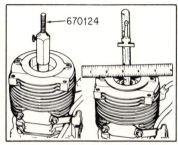

670124

5. INSTALL TIMING TOOL OR RULE

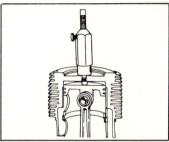

6. FIND TDC (TOP DEAD CENTER)

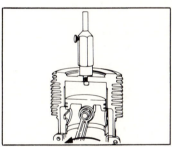

7. BACK OFF ROTATION
(OPPOSITE NORMAL RUNNING
ROTATION)

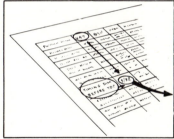

8. FIND BTDC TIMING DIMENSION
(SPECS.)

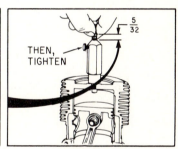

THEN, TIGHTEN

$\frac{5}{32}$

9. APPLY DIMENSION TO TOOL

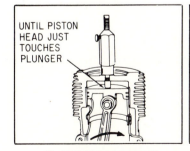

UNTIL PISTON HEAD JUST TOUCHES PLUNGER

10. BRING UP ON STROKE
(NORMAL RUNNING ROTATION)

11. INSTALL TIMING LIGHT
(OR USE CELLOPHANE)

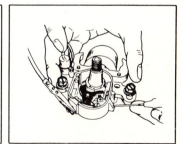

12. ROTATE STATOR UNTIL
POINTS JUST OPEN

FIGURE 14.25 Complete sequence of actions to time one line of two-cycle engines. (*Tecumseh Products Company.*)

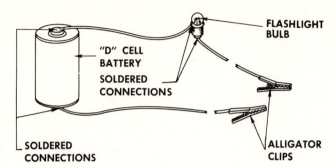

FIGURE 14.26 Timing light made from a flashlight bulb, a dry cell, leads, and two alligator clips.

the flywheel nut finger tight. Now, move the armature and bracket assembly to align the arrows. Remove the flywheel and tighten the armature-bracket mounting screws. Replace the flywheel, with the key, and tighten the nut to the specified tension. Finally, set the armature gap to the proper specifications, as shown in Fig. 14.29, by loosening the two armature attaching screws and moving the armature up or down as necessary. Figure 14.30 shows a simple way to get the proper air gap. Put a postal card between the armature and the flywheel and set the armature down against the postal card. Then tighten the attaching screws and remove the postal card.

14.7 Checking Magneto Parts.   Besides the breaker points, other components of the magneto to be checked are the magnets on the flywheel, the magneto coil, and the condenser.

Figure 14.31 shows one model of coil-condenser tester which can also be used to test automotive-type ignition coils. Some models of tester will also test magneto coils. For a full test, the condenser is

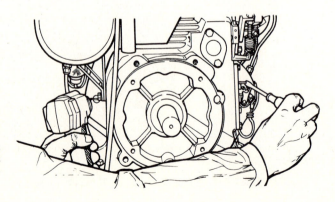

FIGURE 14.27 Using a stroboscopic timing light to time the ignition. (*Kohler Company.*)

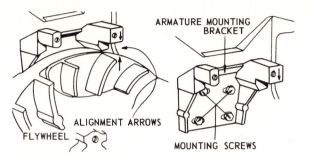

FIGURE 14.28 Timing marks or arrows on the armature mounting bracket and flywheel. (*Briggs and Stratton Corporation.*)

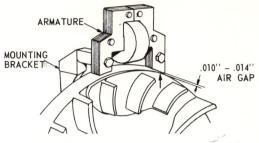

FIGURE 14.29 Correct armature air gap is set, on this model, by shifting the armature up or down as necessary. (*Briggs and Stratton Corporation.*)

FIGURE 14.31 Combination coil and condenser tester. (*Sun Electric Corporation.*)

FIGURE 14.30 A convenient way to set the air gap is to put a postal card which is approximately of the correct thickness, as shown, between the armature and the flywheel.

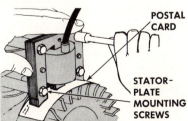

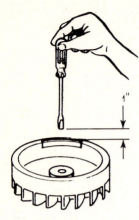

FIGURE 14.32 Testing the strength of the fly-wheel magnets.

checked for capacity, insulation resistance, shorts or grounds, and high series resistance. If the condenser does not meet specifications on any of these tests, it should be replaced.

The magneto coil should be inspected for damage and also tested on an approved coil tester. If it fails to meet specifications, it should be replaced.

The magnets on the flywheel can be tested to see whether or not they are still strong enough to produce adequate magnetism. One method of test is to lay the flywheel on a flat, wooden surface and dangle a screwdriver about an inch above it, as shown in Fig. 14.32. The magnets should strongly attract the screwdriver blade. If the magnets are weak, they should be replaced or remagnetized on a special magnetizer. Alnico magnets cannot be recharged, but must be replaced with new magnets if the old ones have lost strength. Never store flywheels in nested piles. This can cause the magnets to lose their strength.

14.8 Spark Plug Service.   Spark plugs have a tough job. They take the repeated high-voltage surges that produce the sparks, and the tremendous heat and pressure pulses when ignition takes place. Yet, if given a chance, they will work satisfactorily for many hours. Plugs in small engines should be checked and serviced, or replaced periodically in order to maintain top engine performance. There are spark plug cleaners which will clean the plug with a blast of abrasive sand against the electrodes and porcelain interior.

You can also clean deposits out of the plug with a small-bladed knife, as shown in Fig. 14.33. The main things to watch out for are not to damage the porcelain insulator surrounding the center electrode and that you remove all traces of sand or loosened deposits.

NOTE: Some manufacturers warn against using a sand-blast spark plug cleaner and, in fact, void the warranty on their engine if a sand-blasted spark plug is installed in it. Their reasoning is that it is difficult to remove all traces of sand and that it takes only a little sand in the engine to ruin it.

After cleaning the plug, regap it. That is, measure the gap between the electrodes. Do not use a flat gauge because this would result in too great a gap. See Fig. 14.34. Figure 14.35 shows how to check and adjust the spark plug gap.

If you are having plug trouble, refer to Fig. 14.36 to diagnose the cause. This illustration will give you a clue as to what causes various kinds of plug trouble so you can make corrections.

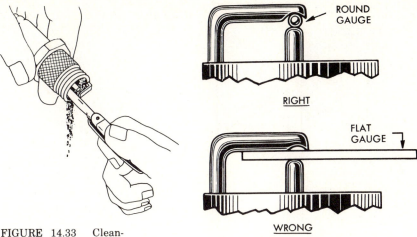

FIGURE 14.33 Cleaning carbon from spark plug shell with a sharp knife.

FIGURE 14.34 Right and wrong gauges to use to check the spark plug gap.

14.9 Battery Ignition Service. In many ways, battery ignition systems are serviced in about the same way as magneto ignition systems. The condenser and coil should be checked if trouble occurs, using a tester such as shown in Fig. 14.31. The breaker points in the distributor are checked, adjusted, and replaced in about the same way as we described for magneto points. Ignition timing is adjusted by loosening the distributor mounting bracket and turning the distributor in its mounting.

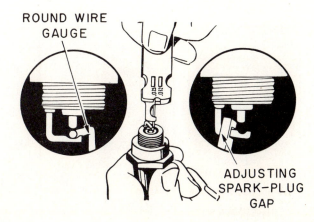

FIGURE 14.35 Using a special gauge and adjusting tool to adjust the spark plug gap.

CARBON FOULED	OIL FOULED	GAP BRIDGED

IDENTIFIED BY BLACK, DRY FLUFFY CARBON DEPOSITS ON INSULATOR TIPS, EXPOSED SHELL SURFACES AND ELECTRODES. CAUSED BY TOO COLD A PLUG, WEAK IGNITION, DIRTY AIR CLEANER, DEFECTIVE FUEL PUMP, TOO RICH A FUEL MIXTURE, IMPROPERLY OPERATING HEAT RISER OR EXCESSIVE IDLING. CAN BE CLEANED.	IDENTIFIED BY WET BLACK DEPOSITS ON THE INSULATOR SHELL BORE ELECTRODES CAUSED BY EXCESSIVE OIL ENTERING COMBUSTION CHAMBER THROUGH WORN RINGS AND PISTONS, EXCESSIVE CLEARANCE BETWEEN VALVE GUIDES AND STEMS, OR WORN OR LOOSE BEARINGS. CAN BE CLEANED IF ENGINE IS NOT REPAIRED, USE A HOTTER PLUG.	IDENTIFIED BY DEPOSIT BUILD-UP CLOSING GAP BETWEEN ELECTRODES. CAUSED BY OIL OR CARBON FOULING. IF DEPOSITS ARE NOT EXCESSIVE, THE PLUG CAN BE CLEANED.
LEAD FOULED	**NORMAL**	**WORN**

IDENTIFIED BY DARK GRAY, BLACK, YELLOW OR TAN DEPOSITS OR A FUSED GLAZED COATING ON THE INSULATOR TIP. CAUSED BY HIGHLY LEADED GASOLINE. CAN BE CLEANED.	IDENTIFIED BY LIGHT TAN OR GRAY DEPOSITS ON THE FIRING TIP. CAN BE CLEANED.	IDENTIFIED BY SEVERELY ERODED OR WORN ELECTRODES. CAUSED BY NORMAL WEAR. SHOULD BE REPLACED.
FUSED SPOT DEPOSIT	**OVERHEATING**	**PRE-IGNITION**
		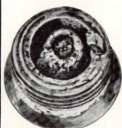
IDENTIFIED BY MELTED OR SPOTTY DEPOSITS RESEMBLING BUBBLES OR BLISTERS. CAUSED BY SUDDEN ACCELERATION. CAN BE CLEANED.	IDENTIFIED BY A WHITE OR LIGHT GRAY INSULATOR WITH SMALL BLACK OR GRAY BROWN SPOTS AND WITH BLUISH-BURNT APPEARANCE OF ELECTRODES, CAUSED BY ENGINE OVERHEATING. WRONG TYPE OF FUEL, LOOSE SPARK PLUGS, TOO HOT A PLUG, LOW FUEL PUMP PRESSURE OR INCORRECT IGNITION TIMING. REPLACE THE PLUG.	IDENTIFIED BY MELTED ELECTRODES AND POSSIBLE BLISTERED INSULATOR. METALLIC DEPOSITS ON INSULATOR INDICATE ENGINE DAMAGE. CAUSED BY WRONG TYPE OF FUEL, INCORRECT IGNITION TIMING OR ADVANCE, TOO HOT A PLUG, BURNT VALVES OR ENGINE OVERHEATING. REPLACE THE PLUG.

FIGURE 14.36   Appearance of spark plugs related to causes of the condition. (*Ford Motor Company.*)

Since small engines do not commonly use battery ignition, and since this is the universal automotive arrangement, we refer you to a book in the McGraw-Hill Automotive Technology Series, "Automotive Electrical Equipment," for further details on how to service battery ignition systems.

CHECKUP

In the chapter you have just completed you learned the fundamentals of small-engine ignition service, how to service magnetos, adjust breaker points, check magneto parts, adjust ignition timing, and service spark plugs. The following questions will not only give you a chance to check up on how well you understand and remember these fundamentals, but also will help you to remember them better. The act of writing down the answers to the questions will fix the facts more firmly in your mind.

NOTE: **Write down your answers in your notebook. Then later you will find your notebook filled with valuable information which you can refer to quickly.**

Completing the Sentences: Test 14.   The sentences below are not complete. After each sentence there are several words or phrases, only one of which will correctly complete the sentence. Write each sentence in your notebook, selecting the proper word or phrase to complete it correctly.

1. One possible cause of engine backfire is (*a*) breaker-point gap too small; (*b*) breaker-point gap too wide; (*c*) defective condenser.
2. Two ways to adjust ignition timing on small engines is to change the breaker-plate position, and (*a*) reposition the flywheel; (*b*) reposition the armature stator; (*c*) readjust breaker-point opening.
3. To check the breaker points of a flywheel magneto, you must first (*a*) remove the spark plug; (*b*) remove the flywheel; (*c*) detach the magneto from the flywheel.
4. The purpose of the dust cover found on many magnetos is to protect the (*a*) breaker points; (*b*) stator windings; (*c*) magnets.
5. To clean breaker points, use (*a*) emery cloth; (*b*) sandpaper; (*c*) a point file.
6. To time the ignition means to make an adjustment that will cause the spark to occur (*a*) just before TDC: (*b*) just after TDC; (*c*) just before BDC.

7. Weak Alnico magnets should be (*a*) recharged; (*b*) replaced; (*c*) heated to restore magnetism.
8. Spark plug electrode gap should be checked with a (*a*) round gauge; (*b*) flat gauge; (*c*) flat feeler.

Written Checkup

In the following, you are asked to write down, in your notebook, the answers to the questions asked or to define certain terms. Writing the answers down will help you to remember them.

1. List the things to check if you do not get an adequate spark on the spark test.
2. Describe in detail a typical flywheel removal procedure.
3. Explain how to check breaker points for wear, damage, and gap.
4. Describe the procedures for adjusting breaker points on different magnetos.
5. Explain how to time the ignition and two ways to check the timing.
6. Explain how to make a timing light.
7. Explain how to use a stroboscopic timing light.
8. Describe spark plug service.

Fuel System Service

15

We described, in an earlier chapter, various types of fuel systems found on small engines, and learned that there are many kinds. These include gravity-feed, suction-feed, and pressure-feed. Complete service of the fuel system includes service to the fuel tank, fuel filter and sediment bowl, carburetor air cleaner, crankcase breather, fuel pump, and governor. We discussed servicing of some of these components in Chap. 11, including fuel filter, sediment bowl, air cleaner, and crankcase breather. Now, in this chapter, we will describe servicing of the fuel tank, fuel pump, carburetor, and governor.

15.1 Fuel-tank Service. Fuel tanks come in a variety of sizes and shapes. There is very little in the way of service that they require. All tanks have a vent of some sort to admit air when fuel is taken out. As a rule, the vent is in the fuel cap. Once in a great while the vent might get clogged up, and this would prevent air from entering so that fuel could not flow out. The result would be that the engine would starve for fuel and stop running.

Some fuel tanks have a cap-and-gauge combination, as shown in Fig. 15.1. The gauge is made of a float on a twisted blade fastened to the indicating needle. The float moves down as the fuel tank loses fuel, and this twists the blade so that the indicating needle moves to indicate the lowered fuel level in the tank.

If the fuel tank is damaged in any way, it should be replaced with a new tank. The tank normally has a bracket which attaches to the engine by screws. Disconnecting the fuel line and removing the screws permits removal of the tank.

15.2 Fuel-pump Service. Most fuel pumps are serviced by complete replacement. They are relatively cheap and it may cost more in labor to repair an old pump than to buy a new one. You can check a fuel pump to see if it works by disconnecting the spark plug wire and the fuel line at the carburetor, and cranking the engine. Hold a small container under the fuel line to catch any fuel that appears. If fuel

FIGURE 15.1 Combination fuel gauge and fuel-tank cap.

flows out strongly and in regular squirts, the fuel pump is working okay. If fuel flow is weak or erratic, there is something wrong with the fuel pump and it should be replaced.

To remove the old pump, disconnect the fuel lines and take out the screws holding the fuel pump on the engine. Lift the fuel pump off, observing the position of the rocker arm (above or below the eccentric?). Install the new pump, making sure the rocker arm goes on the correct side of the eccentric, attach the fuel lines and then tighten the attaching screws. Figure 15.2 shows the installation of a fuel pump on an engine. Note that the pump lever fits into an eccentric groove on the crankshaft and that the part of the lever that rides in the groove should be greased when the pump is installed.

Some manufacturers supply fuel-pump repair information and repair kits. For example, Fig. 15.3 is taken from the service manual of one engine manufacturer, and it shows a disassembled fuel pump used on some of this manufacturer's engines. There is nothing special about disassembling this fuel pump. You should mark the pump cover and pump body with a file as shown so that, on reassembly, you can match the marks and therefore not reverse the cover as it goes on the body.

15.3 Servicing Carburetors. We saw, in earlier chapters, how the carburetor has the job of mixing air and gasoline vapor to provide the engine with the proper air-fuel ratio for good engine operation. If the carburetor is properly adjusted to give this correct air-fuel ratio, it is not likely to go too far out of adjustment in normal operation. However, screws can loosen and throw the adjustment off. In addition, fuel lines and jets in the carburetor can clog. This can mean a partial dis-

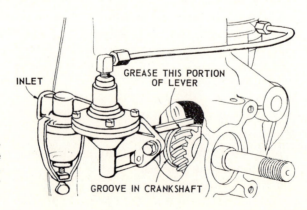

FIGURE 15.2 Installation of fuel pump on one model of small engine. (*Briggs and Stratton Corporation.*)

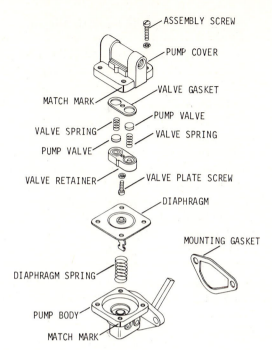

ASSEMBLY SCREW

PUMP COVER

VALVE GASKET

MATCH MARK

PUMP VALVE

VALVE SPRING

VALVE SPRING

PUMP VALVE

VALVE RETAINER

VALVE PLATE SCREW

DIAPHRAGM

MOUNTING GASKET

DIAPHRAGM SPRING

PUMP BODY

MATCH MARK

FIGURE 15.3 Disassembled view of a fuel pump for a small engine. (*Kohler Company.*)

assembly of the carburetor for cleaning which, in turn, means a carburetor adjustment. Our carburetor-servicing story is divided into two parts, adjustments, and removal and rebuilding.

15.4 Preparing to Adjust Carburetor. To do the job correctly, there are certain preliminary steps you should take.

1. Fill the fuel tank except on engines with suction-feed carburetors. On these, fill the tank only half full. Then, when you adjust the suction-feed carburetor, you will be working with an average air-fuel ratio. If you started with a full fuel tank on the suction-feed carburetor, you would adjust correctly for a full tank. Then, as the tank emptied, the mixture would tend to lean out and might become too lean when the tank is nearly empty. The reason? With a nearly empty tank, the fuel must be lifted farther, and less would flow into the passing air stream in the carburetor.

2. Be sure the throttle and governor linkages are free and can move easily.

3. On four-cycle engines, check the oil level in the crankcase and add oil if necessary.

4.   Clean the fuel strainer or filter, and the air filter, as explained in a previous chapter.

5.   Make sure the fuel-tank cap vent is open. If the vent is clogged, it will prevent normal flow from the tank to the carburetor.

6.   Check the ignition and spark plug, as explained in Sec. 10.3.

15.5 Carburetor Adjustments.   A great variety of carburetors have been used on small engines. Figure 15.4 is a sectional view of one type. In this and several following sections, we will describe in general terms how you adjust carburetors. Then, later in the chapter, we will discuss adjustments and repairs of specific models of carburetor.

Usually, the carburetor has three adjustment screws, one to set the idle speed, one to set the idle mixture, and one to set the high-speed load mixture. You may have difficulty deciding which is the idle-mixture adjustment screw and which is the high-speed-load adjustment screw. See Figs. 15.5 to 15.7. Usually, the idle-mixture adjustment screw is closest to the engine, but this is not always true. If you have any doubts, you can check as follows. Start the engine and operate it at idle speed. Then turn the screw you think might be the idle-mixture screw clockwise, or in toward the closed position. If the engine slows down or stops, you know you have found the idle-mixture screw. If the engine speed changes little or not at all, increase engine

FIGURE 15.5   Location of adjustments on one model of carburetor for a small engine.

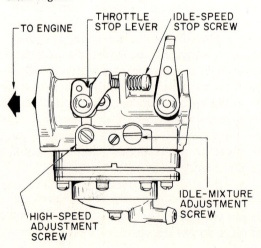

FIGURE 15.4   Sectional view of a carburetor used in a small engine. (*Briggs and Stratton Corporation.*)

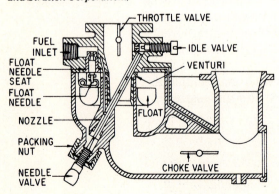

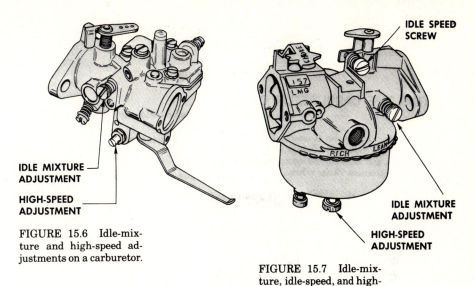

FIGURE 15.6 Idle-mixture and high-speed adjustments on a carburetor.

FIGURE 15.7 Idle-mixture, idle-speed, and high-speed adjustments on a carburetor.

speed to about three-fourths full throttle. Now, if you get a difference in speed as you turn the screw one way or the other, you have found the high-speed load screw.

NOTE: Many carburetors do not have an idle-mixture adjustment. The idle mixture is preset during manufacture and is determined by the size of the discharge port in the carburetor.

15.6 Initial Adjustments. Turn the adjustment screws in until the needles bottom. Then back them off about one turn. This gives an approximate adjustment that should enable you to start the engine and run it until it warms up. Then, the final adjustments can be made in accordance with the manufacturer's specifications. We will describe several specific procedures later in the chapter.

> CAUTION. **Never tighten the adjusting screws more than finger tight. Excessive tightening can cause the needle to jam down into the seat so tightly that both the needle and seat are damaged.**

15.7 Carburetor Repair. If adjustment cannot be made to give good engine operation, then either the carburetor should be replaced with a new one, or else it must be disassembled so that new parts can be installed. There are carburetor repair kits available which contain all necessary new parts. These kits also include all necessary instructions for repairing the carburetors.

> CAUTION. **Always install new gaskets when repairing a carburetor. The old gaskets are probably hardened and will not provide a good seal, so that leakage would occur if they are used again.**

15.8 Choke Adjustment. Most chokes in carburetors for small engines are of the manual type, controlled by linkage to a choke lever. One typical example is shown in Fig. 15.8. To adjust this choke, remove the air cleaner so you can see the choke action. Move the control to the fully closed position. The choke should be closed. If it is not, adjust the control linkage. Then move the control to the fully open position to make sure the choke also moves to the fully open position.

If the choke is controlled by some automatic means such as a thermostat or a solenoid, refer to the manufacturer's service manual for details of the adjustment required. The type with the thermostat can be checked by noting the choke position with the engine cold (it should be partly to fully closed) and with the engine hot (it should be wide open). If the choke does not work this way, then either the thermostat is faulty or else adjustment is required. If the engine uses an electric starter and solenoid, operate the starter and note whether or not the solenoid is actuated. If it is not, then the solenoid is faulty or else the starter switch or circuit is defective.

15.9 High-speed Load Adjustment. This adjustment affects the air-fuel mixture ratio when the engine is operating at rated speed and under full load. The engine must be warmed up to operating

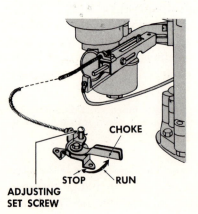

FIGURE 15.8 Adjustable, manually operated choke. (*Briggs and Stratton Corporation.*)

temperature, and it should be under full load when this adjustment is made. If you cannot load the engine to make the adjustment, then make the adjustment without load, but after the adjustment be sure to check the engine operation under normal full load. It is also desirable to use an rpm indicator (a tachometer) to get an accurate reading on engine speed during the adjustment. A relatively simple type of rpm indicator is the vibrotachometer, shown in use in Fig. 15.9. This device is rested on the cowling of the engine and senses the power strokes, each of which produces a definite vibration of the cowling. The device then translates this information into rpm and indicates it on a scale on the side of the testing tool.

Now, with the engine operating at full speed, turn the high-speed load adjustment screw in slowly until the engine begins to slow down. When this happens, the air-fuel mixture has leaned out so much that the engine power is reduced. Now, slowly turn the adjustment screw out until the engine slows down or the exhaust begins to turn black. At this position, the needle is passing too much gasoline so the mixture is over-rich; not all the gasoline burns and thus turns the exhaust black. Next, slowly turn the adjustment screw in until the engine runs smoothly and at full speed.

NOTE: **Make adjustments of about one-eighth turn at a time and wait a few seconds between turns for the engine to adjust to the changed mixture richness.**

15.10 Idle-mixture Adjustment. This adjustment affects the mixture richness when the engine is idling. As previously mentioned, many carburetors do not have this adjustment; the idle-mixture port is fixed. If the carburetor does have the adjustment, it is made with the engine running and warmed up. First, turn the idle-speed adjustment screw, with the engine idling, to get the lowest engine speed possible without stalling. Next, turn the idle-mixture adjustment screw in until

VIBRO-
TACHOMETER

FIGURE 15.9 Using a vibrotachometer, a simple type of rpm indicator.

the engine begins to slow down or roll. This means that the mixture is too lean to support normal engine operation. Now, turn the adjustment screw back out slowly until the engine idles smoothly. Recheck the high-speed load adjustment to make sure it is still okay. Then operate the throttle several times from idle to full speed to make sure the engine will go from idle to full speed and back again without hesitation. Finally, adjust the idle speed as explained in the following paragraph.

15.11 Idle-speed Adjustment.   This adjustment is controlled by a stop screw which can be turned in or out to change the idle speed. Its basic purpose is to prevent the throttle valve from closing completely so that the engine would stall. Small engines usually are designed to idle at fairly high speeds of somewhere from 1,200 to 3,000 rpm (revolutions per minute). Always check the manufacturer's service manual to determine the specified speed before attempting to set the idle speed. You will find that most specifications call for setting the idle speed at about one-half full speed. If the engine is idled too slowly, the mixture richness is apt to be too high with the result that the plugs, pistons, and exhaust ports (on two-cycle engines) will soon foul up from carbon, due to only partly burned gasoline.

Idle speed should be set with the engine warmed up, and the other settings discussed above should be made first. Then, a tachometer should be used to measure the speed while the idle-speed adjustment screw is turned to obtain the specified speed.

15.12 Float Adjustment.   The float should be adjusted so that the proper level of gasoline will be maintained in the float bowl. Normally, this adjustment will not change. However, if the carburetor requires repair, then this adjustment should be checked. The procedure of checking the float level on one model of carburetor is shown in Fig. 15.10. The float should be parallel to the body mounting surface with the body gasket in place and the float valve and float installed. Bend the tang on the float, if necessary, to bring the float to parallel.

SERVICING SPECIFIC CARBURETOR MODELS

15.13 Recommended Adjustments on Specific Models.   Following are instructions, taken from manufacturer's service manuals, on how to service specific models of carburetors. These are examples of what you will find when you work on specific carburetors and study the service manuals that cover them.

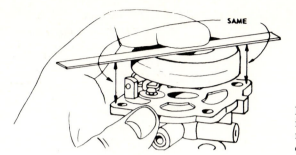

FIGURE 15.10 Checking the float level on one model carburetor. (*Briggs and Stratton Corporation.*)

NOTE: If you have to remove a carburetor from the engine, be sure you study the arrangements of the throttle and governor linkages so you can put everything back in the original positions. If you are not sure you can remember the linkage arrangements, you had better make a pencil sketch which will show where everything goes.

15.14 Lawn Boy D Engine Carburetor Service. Figure 15.11 shows the carburetor used on this engine with the various serviceable and adjustable parts shown separately. The slow-speed needle is adjusted to provide the proper mixture richness on idle, and the main needle is adjusted to provide the proper mixture richness during full-speed, full-load operation.

Figure 15.12 shows the float system in the carburetor. The float level will seldom require adjustment, but if it does, refer to Fig. 15.13 which shows the correct adjustment. To check this adjustment, remove the carburetor from the engine (Fig. 15.14). The carburetor and reed plate must be removed as an assembly. The reed plate is between the carburetor and engine. The reed plate can then be separated from the carburetor.

> CAUTION. **Handle the reed plate with care to avoid damaging the reeds. The reeds must lie flat against the reed plate in order to form a good seal. If the reeds are bent or damaged in any way, replace the complete reed-plate assembly.**

Remove the float bowl and gasket from the carburetor and turn the carburetor upside down. Then, measure the distance between the top of the float and the carburetor body with the float arm resting on the float-valve needle. Adjust, if necessary, by bending the float arm.

NOTE: Carburetor parts can be cleaned in solvent and passages in the carburetor body blown out with compressed air. But do not clean cork floats in solvent because this could dissolve the varnish coat and ruin them. Never wipe carburetor parts with cloth; lint can get into carburetor passages and clog them.

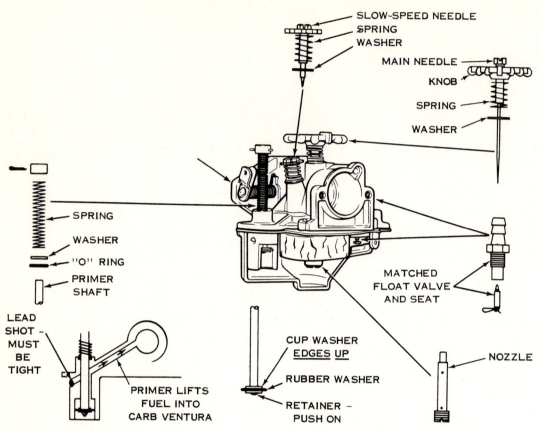

FIGURE 15.11 Disassembled view of a carburetor for a small engine used on a lawn mower. (*Lawn Boy Division, Outboard Marine Corporation.*)

You can follow Fig. 15.11 in disassembling the carburetor to replace worn parts. Note that the float valve and seat come in a matched pair and both must be replaced at the same time. Always use all new gaskets when reassembling the carburetor. Also, when installing the nozzle in the carburetor, unscrew the control knob in order to avoid tightening the nozzle against the needle and damaging both.

15.15 Kohler Small-Engine Carburetor Service. Figure 15.15 and 15.16 show the adjustments on the side-draft and the updraft carburetors used on many models of Kohler small engines. Kohler provides the following trouble-diagnosis chart.

Condition	Possible Cause and Remedy
A. Black, sooty exhaust smoke, engine sluggish	Mixture too rich — readjust main fuel needle
B. Engine misses and backfires at high speed	Mixture too lean — readjust main fuel needle
C. Engine starts, sputters, and dies under cold-weather starting	Mixture too lean — turn main fuel adjustment one-fourth turn counterclockwise
D. Engine runs rough or stalls at idle speed	Improper idle adjustment — readjust idle-fuel needle

Here is the recommended procedure for adjusting the two carburetors shown in Figs. 15.15 and 15.16. With the engine stopped, turn the main-fuel and idle-fuel needles all the way in (clockwise) until they bottom. Do not tighten screws as this will damage the needle valves.

FIGURE 15.12 Sectional view of assembled carburetor shown in previous illustration. (*Lawn Boy Division, Outboard Marine Corporation.*)

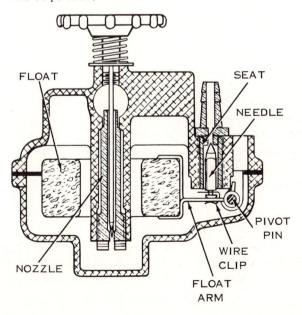

FIGURE 15.13 Correct adjustment for the float. (*Lawn Boy Division, Outboard Marine Corporation.*)

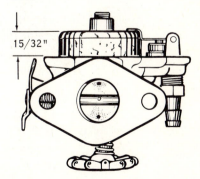

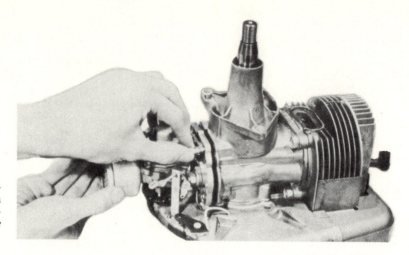

FIGURE 15.14 Removing the carburetor from the engine. (*Lawn Boy Division, Outboard Marine Corporation.*)

Make preliminary adjustments by backing the main-fuel screw out two turns. Back out the idle-fuel screw $1\frac{1}{4}$ turns. Start the engine and operate it at normal speed until it reaches operating temperature. With the engine operating at full throttle and full load, turn the main-

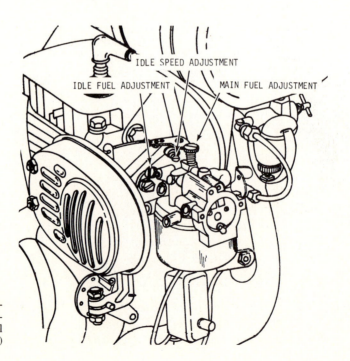

FIGURE 15.15 Adjustments on a side-draft carburetor used on a small engine. (*Kohler Company.*)

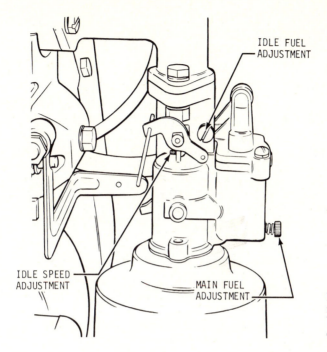

IDLE FUEL
ADJUSTMENT

IDLE SPEED
ADJUSTMENT

MAIN FUEL
ADJUSTMENT

FIGURE 15.16 Adjustments on an updraft carburetor used on a small engine. (*Kohler Company.*)

fuel needle in slowly until the engine loses speed. Note the position of the screw. Turn the needle out until the engine regains speed and then begins to slow down again from overrichness. Then turn the needle back in until it is halfway between the two extreme positions.

To make the idle-fuel adjustment, operate the engine at idle speed of about 1,000 rpm. Adjust the idle-speed screw to get this rpm. Check the speed with a tachometer. Now turn the idle-fuel needle in until the engine slows down and idles roughly. Then turn the screw out until engine speeds up and idles smoothly at the desired idle speed.

Recheck the two adjustments because adjusting the idle-fuel mixture has some effect on the main-fuel adjustment.

If the adjustments cannot be made to achieve good engine operation, and your checks show the trouble is not in the ignition system or engine itself, you will need to remove the carburetor for service. Figures 15.17 and 15.18 are disassembled views of the side-draft and updraft carburetors. When disassembling a carburetor, use the repair kit for that model carburetor when reassembling it. The kit has replacement parts for everything that might wear in the carburetor. Always use the new gaskets. Never try to reuse old gaskets; they will

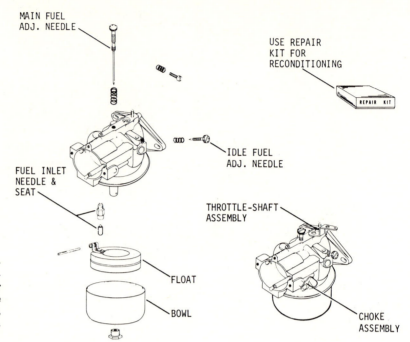

FIGURE 15.17 Left, disassembled view of sidedraft carburetor. Upper right, repair kit for the carburetor. Lower right, assembled carburetor. (*Kohler Company.*)

probably allow leakage. Follow the illustrations when disassembling the carburetors. Do not try to remove the choke and throttle plates and shafts. If these parts are worn, discard the old carburetor.

Clean the carburetor parts in solvent and blow out passages in the carburetor body with compressed air. Never wipe the carburetor parts with cloths; lint could get into carburetor passages and clog them. Examine all parts for wear, and if parts are worn, get a repair kit for the carburetor which will have all the necessary parts to rebuild it.

1. Assembly of Side-draft Carburetor. Install the needle seat, needle, and float, and float pin. Adjust the float level with the carburetor inverted and the float resting lightly against the needle. There should be $\frac{11}{64}$ in. clearance between the machined surface of the carburetor and the free end of the float (the end opposite the needle valve). Adjust by bending the lip of the float with a small screwdriver. Then install other parts. Adjust the main-fuel and idle-fuel adjustment screws, as already explained.

2. Assembly of Updraft Carburetor. Install the needle seat with a $\frac{5}{16}$-in. socket and torque wrench. Torque to 25–30 in.-lb—no more.

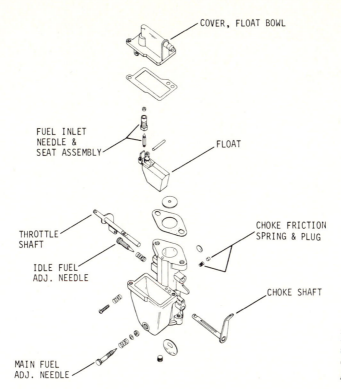

COVER, FLOAT BOWL

FUEL INLET
NEEDLE &
SEAT ASSEMBLY

FLOAT

THROTTLE
SHAFT

CHOKE FRICTION
SPRING & PLUG

IDLE FUEL
ADJ. NEEDLE

CHOKE SHAFT

MAIN FUEL
ADJ. NEEDLE

FIGURE 15.18 Disassembled view of updraft carburetor. (*Kohler Company.*)

Overtightening will ruin it. Install the needle, float, and float pin. Set the float level with the bowl cover casting inverted and the float resting lightly against the needle in its seat. There should be $\frac{7}{16}$ in. clearance between the machined surface of the casting and the free end of the float (the side opposite the needle valve). Adjust by bending the lip of the float with a small screwdriver. Install the other parts and adjust the main-fuel and idle-fuel adjustment screws, as already explained. Install the idle-speed screw and spring, put the carburetor on the engine, and adjust it to get the idle speed desired.

15.16 Tecumseh Carburetor Service. Two models of carburetors used on Tecumseh small engines (Lauson and Power Products) are shown in Figs. 15.19 and 15.20. The diaphragm carburetor (Fig. 15.20) is used on power saws where the carburetor must work regardless of the position in which it is operating. Figures 15.21 to 15.23 illustrate the operation of this carburetor during choking when the engine is started, when the engine is idling, and when the engine is operating at high speed.

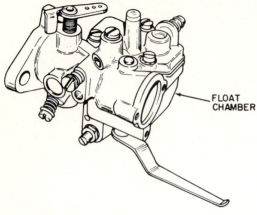

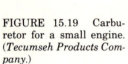

FIGURE 15.19 Carburetor for a small engine. (*Tecumseh Products Company.*)

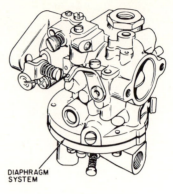

FIGURE 15.20 Diaphragm carburetor for a power saw. (*Tecumseh Products Company.*)

The adjustments of these carburetors are very similar to the adjustments of the carburetors we have covered previously. Approximate settings on the main-mixture and idle-mixture screws is one turn back from fully seated needles. The idle-speed screw (on top of the carburetor) is adjusted by backing out the screw and then turning it in until the screw just touches the throttle lever. Turn it in one more turn.

FIGURE 15.21 Operation of diaphragm carburetor during starting, when choke is closed. (*Tecumseh Products Company.*)

FIGURE 15.22 Operation of diaphragm carburetor during idle. (*Tecumseh Products Company.*)

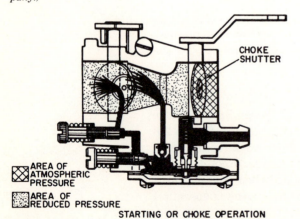

STARTING OR CHOKE OPERATION

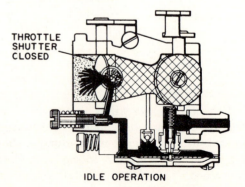

IDLE OPERATION

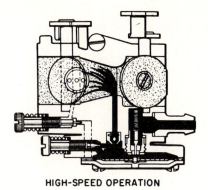

HIGH-SPEED OPERATION

FIGURE 15.23 Operation of diaphragm carburetor during high-speed running. (*Tecumseh Products Company.*)

If the carburetors require service, they must be removed from the engine and disassembled. Refer to Figs. 15.24 and 15.25 for details of how to service the two carburetors.

Figure 15.26 shows another float-type carburetor, and Fig. 15.27 shows the float and needle arrangement in this carburetor. Figure 15.28 shows how to check the float setting. If the setting is not correct, remove the float to make the adjustment. Adjustment is made by bending the tab on the float hinge.

15.17 Servicing Briggs and Stratton Carburetors. These carburetors are serviced in about the same way as carburetors we have previously discussed. Figure 15.29 shows the adjustments to be made on one model. Idle and main needle valves are adjusted by turning them all the way in, and then backing them off the specified amount for the preliminary adjustment. Then final adjustments are made to get the best operating conditions on idle and full throttle. Figure 15.30 shows another carburetor in which the adjusting needle valves are differently located. Adjustment procedures, however, are very similar.

The carburetor shown in Fig. 15.31 is used on such machines as lawn mowers. It operates from a single control lever, as shown in Fig. 15.32, which allows the operator to choke, vary engine speed, and stop the engine from a single control. Adjustments are as follows.

Choke Operation. Remove air cleaner. Move remote control lever to CHOKE position. The carburetor choke should then be closed. Move the remote control lever to STOP. Control lever on carburetor should then make good contact with stop switch to short out ignition.

FIGURE 15.24 Details of how to service the carburetor shown in Fig. 15.19 (*Tecumseh Products Company.*)

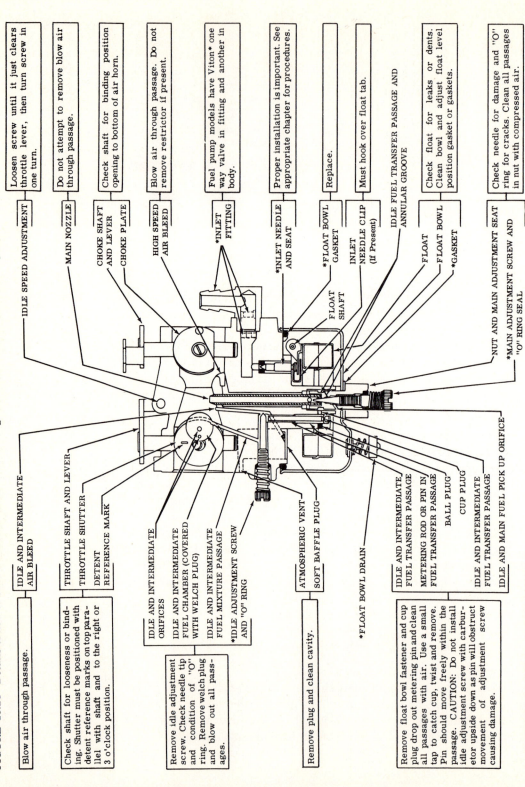

Blow air through passage.

IDLE AND INTERMEDIATE AIR BLEED

Loosen screw until it just clears throttle lever, then turn screw in one turn.

IDLE SPEED ADJUSTMENT

Do not attempt to remove blow air through passage.

MAIN NOZZLE

Check shaft for binding position opening to bottom of air horn.

CHOKE SHAFT AND LEVER

CHOKE PLATE

Blow air through passage. Do not remove restrictor if present.

HIGH SPEED AIR BLEED

Fuel pump models have Viton* one way valve in fitting and another in body.

*INLET FITTING

Proper installation is important. See appropriate chapter for procedures.

*INLET NEEDLE AND SEAT

Replace.

*FLOAT BOWL GASKET

Must hook over float tab.

INLET NEEDLE CLIP (If Present)

FLOAT SHAFT

IDLE FUEL TRANSFER PASSAGE AND ANNULAR GROOVE

FLOAT

FLOAT BOWL

Check float for leaks or dents. Clean bowl and adjust float level position gasket and gaskets.

*GASKET

Check needle for damage and "O" ring for cracks. Clean all passages in nut with compressed air.

NUT AND MAIN ADJUSTMENT SEAT

*MAIN ADJUSTMENT SCREW AND "O" RING SEAL

Check shaft for looseness or binding. Shutter must be positioned with detent reference marks on top parallel with shaft and to the right or 3 o'clock position.

THROTTLE SHAFT AND LEVER

THROTTLE SHUTTER

DETENT REFERENCE MARK

Remove idle adjustment screw. Check needle tip and condition of "O" ring. Remove welch plug and blow out all passages.

IDLE AND INTERMEDIATE ORIFICES

IDLE AND INTERMEDIATE FUEL CHAMBER (COVERED WITH WELCH PLUG)

IDLE AND INTERMEDIATE FUEL MIXTURE PASSAGE

*IDLE ADJUSTMENT SCREW AND "O" RING

Remove plug and clean cavity.

ATMOSPHERIC VENT

SOFT BAFFLE PLUG

*FLOAT BOWL DRAIN

Remove float bowl fastener and cup plug drop out metering pin and clean all passages with air. Use a small tap to catch cup, twist and remove. Pin should move freely within the passage. CAUTION: Do not install idle adjustment screw with carburetor upside down as pin will obstruct movement of adjustment screw causing damage.

IDLE AND INTERMEDIATE FUEL TRANSFER PASSAGE

METERING ROD OR PIN IN FUEL TRANSFER PASSAGE

BALL PLUG

CUP PLUG

IDLE AND INTERMEDIATE FUEL TRANSFER PASSAGE

IDLE AND MAIN FUEL PICK UP ORIFICE

* NON METALLIC ITEMS - CAN BE DAMAGED
BY HARSH CARBURETOR CLEANERS

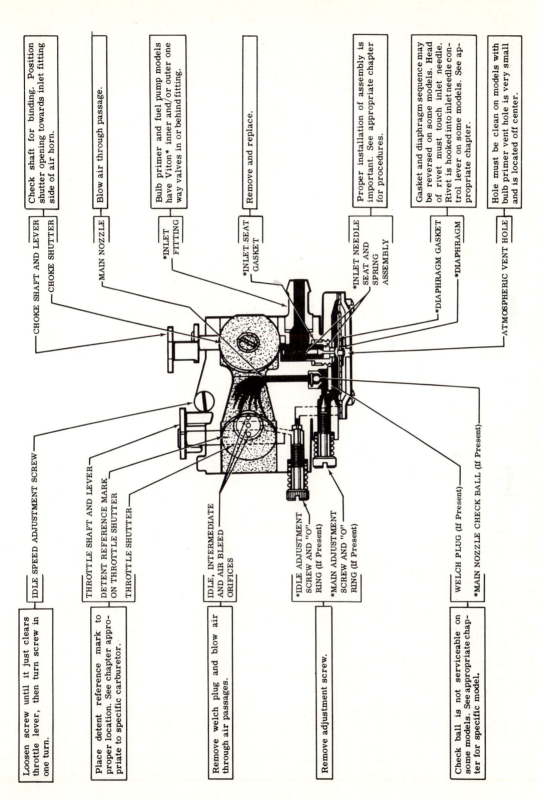

Check shaft for binding. Position shutter opening towards inlet fitting side of air horn.

CHOKE SHAFT AND LEVER

CHOKE SHUTTER

Blow air through passage.

MAIN NOZZLE

Bulb primer and fuel pump models have Viton* inner and/or outer one way valves in or behind fitting.

*INLET FITTING

Remove and replace.

*INLET SEAT GASKET

Proper installation of assembly is important. See appropriate chapter for procedures.

*INLET NEEDLE SEAT AND SPRING ASSEMBLY

Gasket and diaphragm sequence may be reversed on some models. Head of rivet must touch inlet needle. Rivet is hooked into inlet needle control lever on some models. See appropriate chapter.

*DIAPHRAGM GASKET

*DIAPHRAGM

Hole must be clean on models with bulb primer vent hole is very small and is located off center.

ATMOSPHERIC VENT HOLE

Loosen screw until it just clears throttle lever, then turn screw in one turn.

IDLE SPEED ADJUSTMENT SCREW

Place detent reference mark to proper location. See chapter appropriate to specific carburetor.

THROTTLE SHAFT AND LEVER

DETENT REFERENCE MARK ON THROTTLE SHUTTER

THROTTLE SHUTTER

Remove welch plug and blow air through air passages.

IDLE, INTERMEDIATE AND AIR BLEED ORIFICES

Remove adjustment screw.

*IDLE ADJUSTMENT SCREW AND "O" RING (If Present)

*MAIN ADJUSTMENT SCREW AND "O" RING (If Present)

Check ball is not serviceable on some models. See appropriate chapter for specific model.

WELCH PLUG (If Present)

*MAIN NOZZLE CHECK BALL (If Present)

* NON METALLIC ITEMS – CAN BE DAMAGED BY HARSH CARBURETOR CLEANERS

FIGURE 15.25   Details of how to service the diaphragm carburetor shown in Fig. 15.20. (*Tecumseh Products Company.*)

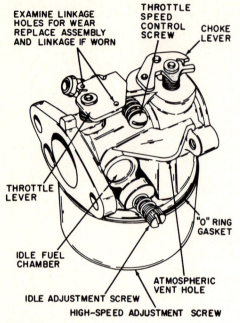

EXAMINE LINKAGE HOLES FOR WEAR REPLACE ASSEMBLY AND LINKAGE IF WORN

THROTTLE SPEED CONTROL SCREW

CHOKE LEVER

THROTTLE LEVER

IDLE FUEL CHAMBER

"O" RING GASKET

ATMOSPHERIC VENT HOLE

IDLE ADJUSTMENT SCREW

HIGH-SPEED ADJUSTMENT SCREW

FIGURE 15.26 Adjustments on a carburetor using a float-feed system. (*Tecumseh Products Company.*)

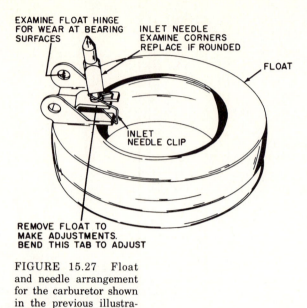

EXAMINE FLOAT HINGE FOR WEAR AT BEARING SURFACES

INLET NEEDLE EXAMINE CORNERS REPLACE IF ROUNDED

FLOAT

INLET NEEDLE CLIP

REMOVE FLOAT TO MAKE ADJUSTMENTS. BEND THIS TAB TO ADJUST

FIGURE 15.27 Float and needle arrangement for the carburetor shown in the previous illustration. (*Tecumseh Products Company.*)

FIGURE 15.28 Using a drill to check the float-level adjustment. (*Tecumseh Products Company.*)

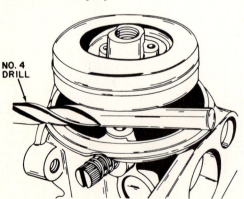

NO. 4 DRILL

FIGURE 15.29 Final adjustments on one model of carburetor. (*Briggs and Stratton Corporation.*)

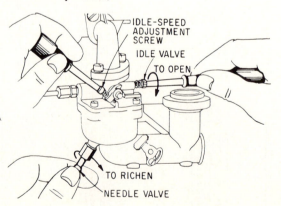

IDLE-SPEED ADJUSTMENT SCREW

IDLE VALVE

TO OPEN

TO RICHEN

NEEDLE VALVE

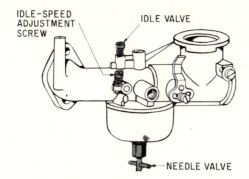

FIGURE 15.30 Adjusting valves and screw on a carburetor. (*Briggs and Stratton Corporation.*)

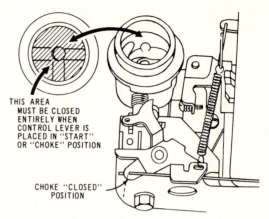

FIGURE 15.31 Carburetor used on a lawn mower with a single control lever. (*Briggs and Stratton Corporation.*)

To adjust, see Fig. 15.33. Place remote control lever on equipment in FAST high-speed position. Lever *C* on carburetor should be just touching choke arm at *D*. To adjust, loose casing clamp screw *A* on blower housing. Move control casing *B* forward or backward until correct position is obtained. Tighten screw *A*.

Recheck operation of controls after adjustment. Replace air cleaner.

FIGURE 15.32 Lawn mower with a single control lever. (*Briggs and Stratton Corporation.*)

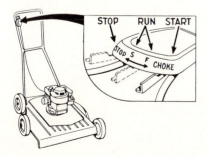

FIGURE 15.33 Adjustment of linkage to choke arm. (*Briggs and Stratton Corporation.*)

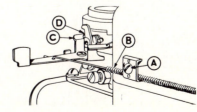

Manual Control Check. Move control lever to left as far as possible. Choke should be fully closed. Now move control lever to right as far as possible. Stop switch blade should make contact with control lever. See Fig. 15.34.

If choke does not close completely or if stop switch does not make contact, control plate must be readjusted.

Adjusting Control Plate. See Fig. 15.35. Move lever to left until it snaps into run detent. Arm *B* should just touch choke arm at *C*. If it does not, loosen screws *A* slightly and move control plate to right or left until lever just touches choke arm at *C*. Tighten screws *A*.

Initial Adjustment. Turn needle valve clockwise to close it. Then open two turns. This initial adjustment will permit the engine to be started and warmed up before making final adjustment.

Final Adjustment. See Fig. 15.36. With engine running at normal operating speed (approximately 3,000 rpm without load) turn needle valve clockwise until engine starts to lose speed (lean mixture). Then slowly turn needle valve counterclockwise past the point of smoothest operation, until engine just begins to run unevenly. This mixture will give best performance under load.

Hold throttle in idling position. Turn idle-speed adjusting screw until fast idle is obtained (1,750 rpm).

Test the engine under full load. If engine tends to stall or die out, it usually indicates that the mixture is slightly lean and it may be necessary to open the needle valve slightly to provide a richer mixture. This richer mixture may cause a slight unevenness in idling.

FIGURE 15.34 Correct positions for lever when choking or in the stopped position. (*Briggs and Stratton Corporation.*)

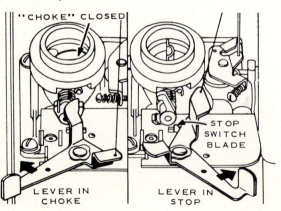

FIGURE 15.35 Adjusting control plate. (*Briggs and Stratton Corporation.*)

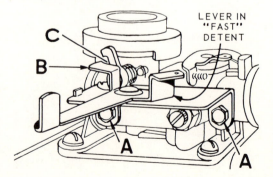

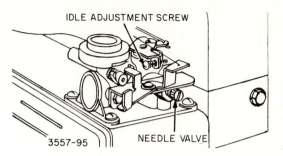

IDLE ADJUSTMENT SCREW

3557-95

NEEDLE VALVE

FIGURE 15.36 Locations of needle valve and idle adjusting screw. (*Briggs and Stratton Corporation.*)

GOVERNOR SERVICE

15.18 Governor Service. We will now discuss the procedures for checking and servicing governors. As you will recall from Chap. 8, there are two general types of governor, the air-vane type and the centrifugal type. The air-vane type works on a blast of air from the blades on the engine flywheel. The centrifugal type works off a centrifugal device which is actuated by engine speed. If you are not clear about how these two types of governor operate, refer to Secs. 8.18 to 8.21.

If the engine is not governed at the correct speed, the governor should be adjusted. On some models, this can be done by bending the link between the throttle and the governor or making some similar linkage adjustment. On others, the spring can be changed to change the engine speed (see Fig. 15.37). Do not attempt to adjust the governor speed by stretching the spring. This will not work because you may weaken the spring and it will probably go back to its original set after a while. There are several other adjustment methods as we will explain in following paragraphs.

15.19 Adjusting Governors. Figure 15.38 shows various methods of adjusting governors of different designs. One of the most important things to remember when adjusting a governor is not to increase

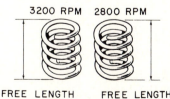

FIGURE 15.37 On some engines, the governor spring must be changed to change governed engine speed. (*Lawn Boy Division, Outboard Marine Corporation.*)

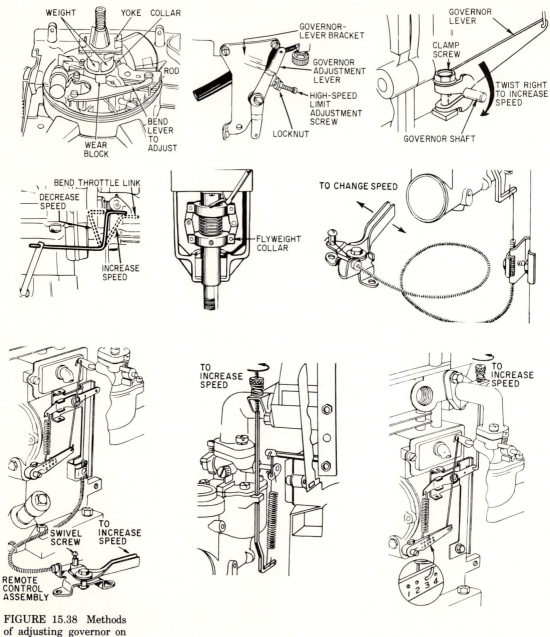

FIGURE 15.38 Methods of adjusting governor on different engines. The nine methods shown here include most adjustment methods.

engine speed beyond the specified value. Excessive engine speed greatly shortens engine life.

If you cannot find the design of governor linkage you are working on in Fig. 15.38, refer to the service manual covering the model engine on which the governor is mounted.

15.20 Servicing Governors. From the standpoint of governor service, you will find three basic types, the air-vane type, the internally mounted centrifugal type, and the externally mounted centrifugal type.

1. Vane-type Governor. There is little in the way of service this governor requires. To get to the governor, remove the engine shroud. Make sure the vane is not bent and that it is free to move when air blows on it.

2. Externally Mounted Centrifugal Governor. This type of governor, mounted under the flywheel, as shown in Fig. 8.39, requires removal of the flywheel, as previously explained in Chap. 14. If you are going to take the governor apart, be sure you notice just how the parts go together so you will have no trouble reinstalling the governor.

3. Internally Mounted Centrifugal Governor. If the governor is mounted inside the engine crankcase, as shown in Fig. 15.39, you will have to disassemble the engine to get to the governor. These governors are often somewhat more complicated in design, so you should be especially careful to note the relationship of all parts before attempting to disassemble the governor.

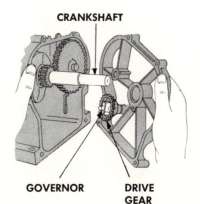

CRANKSHAFT

GOVERNOR   DRIVE GEAR

FIGURE 15.39 Location of an internally mounted governor on one engine.

CHECKUP

You learned, in the chapter you have just completed, how to service small-engine fuel systems, including the fundamentals of servicing different kinds of carburetors. The following questions will not only give you a chance to check up on how well you understand and remember these fundamentals, but also will help you to remember them better. The act of writing down the answers to the questions will fix the facts more firmly in your mind.

NOTE: Write down your answers in your notebook. Then later you will find your notebook filled with valuable information which you can refer to quickly.

Completing the Sentence: Test 15.   The sentences below are not complete. After each sentence there are several words or phrases, only one of which will correctly complete the sentence. Write each sentence in your notebook, selecting the proper word or phrase to complete it correctly.

1. Most fuel pumps for small engines are serviced by (*a*) rebuilding; (*b*) replacement; (*c*) replacement of valves.
2. The three adjustment screws found on many carburetors for small engines are to adjust (*a*) idle speed, low speed, high speed; (*b*) idle speed, idle mixture, governed speed; (*c*) idle speed, idle mixture, high-speed load mixture.
3. Turning the high-speed adjustment screw (*a*) enriches the mixture; (*b*) leans out the mixture; (*c*) increases engine speed.
4. The float on most carburetors is adjusted by (*a*) bending a tang; (*b*) turning a screw; (*c*) replacing the float assembly.
5. To make the float adjustment on the Lawn Boy D engine carburetor, you first (*a*) remove the reed plate; (*b*) remove the carburetor; (*c*) adjust the high-speed load mixture screw.
6. If the engine starts, sputters, and dies under cold-weather starting, chances are the (*a*) ignition system is weak; (*b*) mixture is too lean; (*c*) mixture is too rich.
7. The approximate preliminary settings of the high-speed and idle mixture screws is (*a*) tight against the seat; (*b*) back from seating about three turns; (*c*) back from seating about one to two turns.
8. The two basic types of governor for small engines are (*a*) air-vane and centrifugal; (*b*) internal and external; (*c*) air-vane and diaphragm.

Written Checkup

In the following, you are asked to write down, in your notebook, the answers to the questions asked or to define certain terms. Writing the answers down will help you to remember them.

1. Explain how to remove and install a fuel pump.
2. Make a list of the preliminary steps you must take before you adjust a carburetor.
3. Describe a typical carburetor adjustment of the type using three adjustment screws. Of the type having a single control lever.
4. Make a list of the troubles and possible causes that Kohler listed in its fuel-system trouble-diagnosis chart.
5. Describe a typical carburetor disassembly-reassembly procedure.
6. List different governor-adjustment procedures.

Servicing Small Engines

16

A great variety of small-engine designs will be found in small powered machinery. Thus, a great variety of servicing procedures are required. However, the fundamentals are all similar. That is, magneto contact point adjustments, carburetor adjustments, engine disassembly, bearing replacement, cylinder honing, and so on, are all quite similar for the different engines. In this chapter, we will cover these fundamentals as they apply to a two-cycle and a four-cycle engine.

16.1 Eliminating Trouble. When trouble occurs in an engine, eliminating the trouble may require anything from a very minor adjustment to a complete disassembly of the engine so that worn or defective parts can be serviced or replaced. Sometimes all that is required is an adjustment to the carburetor or governor, or replacement of the contact points in the magneto. At other times, a complete overhaul of the engine is required, with the engine being completely disassembled so the individual parts can be checked and then serviced or replaced with new parts. A previous chapter discussed various engine troubles, their possible causes, and the service or adjustments that might be needed to eliminate these troubles.

NOTE: These engines are relatively inexpensive and repair time and parts are relatively costly. Thus, if an engine is in bad shape, and a new piston, rings, connecting rod, and bearings are required, plus honing of the cylinder walls, all this probably would cost more than a new engine. In such case, the best thing to do would be to throw away the old engine and buy a new one. However, the old engine may be used by students to practice on, giving them experience in disassembling and reassembling it, honing the cylinder walls, and so on.

Many engine manufacturers supply "short-block" engines. These are engines that do not have such add-on parts as the magneto, cylinder head, carburetor, and starter. These short-block engines include the cylinder block, piston with rings, connecting rod, crankshaft, and, on four-cycle engines, the valves and camshaft. They come from the factory with all new parts, ready for you to install the rest of the parts needed to complete the engine. The short-block engine is much less expensive than the complete engine and is worth considering if an engine you are repairing needs a lot of internal work such as honing cylinders, replacing bearings, and so on.

TWO-CYCLE ENGINE SERVICE

The following paragraphs describe the disassembly, servicing, and reassembly of a two-cycle engine used in a lawn mower. Figure 16.1 is a cutaway view of the engine.

16.2 Engine Disassembly.   As a first step, clean the engine as explained in Chap. 11. Then disconnect the gas line from the carburetor, remove the choke rod from the choke shaft, take out the attaching screws, and remove the complete engine shroud, gasoline tank, and recoil starter as an assembly. See Fig. 16.2.

Next, remove the flywheel. We described various methods of doing this in Chap. 14. Here is the way Lawn Boy recommends doing the job. Remove the flywheel nut and washer. To hold the flywheel while the nut is being loosened, a piston stop can be used on some models. See Fig. 16.3. This device is installed in place of the spark plug and has a long rod which comes into contact with the piston head. Now, when the flywheel nut is turned, the piston moves up to the stop and the stop keeps the piston, crankshaft, and flywheel from moving.

Lift the starter pulley, plate, screen, pin, and spring off the flywheel. See Fig. 16.4. Remove the flywheel. If it sticks, put the nut back on the end of the crankshaft, turning it down until it is nearly flush with the end of the crankshaft. Now, use a soft metal hammer, lift up on the flywheel, and tap gently on the nut. See Fig. 16.5. The nut pro-

FIGURE 16.1 Cutaway view of the two-cycle engine in a lawn mower. (*Lawn Boy Division, Outboard Marine Corporation.*)

FIGURE 16.2 Removing engine shroud, gas tank, and starter as an assembly. (*Lawn Boy Division, Outboard Marine Corporation.*)

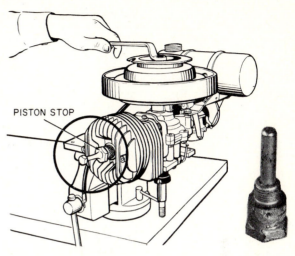

FIGURE 16.3 Using piston stop to hold the piston while flywheel nut is being removed. The piston stop is shown to the right. (*Lawn Boy Division, Outboard Marine Corporation.*)

FIGURE 16.4 Removing starter pulley, plate, and screen. (*Lawn Boy Division, Outboard Marine Corporation.*)

FIGURE 16.5 Hammering on crankshaft nut to loosen the flywheel. (*Lawn Boy Division, Outboard Marine Corporation.*)

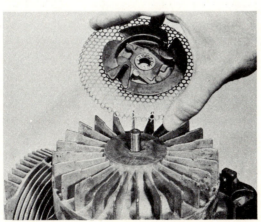

tects the end of the crankshaft so the screw threads will not become battered. Note that there are other ways to remove the flywheel, and various flywheel pullers, as explained in Sec. 14.3.

Lift off the governor yoke, arms, and collar as an assembly and set it to one side. See Fig. 16.6. Remove the governor lever and wear-block assembly. See Fig. 16.7. Remove three screws and lift off the magneto plate. See Fig. 16.8.

Take out the screws and remove the carburetor, air filter, and reed-plate assembly as a unit. See Fig. 15.14. Bend down the tangs of the lock plates and remove the two screws attaching the connecting-rod cap to the rod using either a screwdriver or a box wrench as required by the type of screw. See Fig. 16.9.

Remove the cylinder bolts and separate the crankcase from the cylinder. See Fig. 16.10. Now, you can slip the piston-and-rod assembly out of the cylinder, as shown in Fig. 16.11.

Before you slip the piston clear out of the cylinder, move it back and forth a few times to check for binding. If the piston moves hard in the cylinder, then there is probably a scuffed condition that will require replacement of the piston and rings and honing of the cylinder wall. Examine the piston, after removing it, for scratches, scuff marks, or scores. Any of these indicate undue wear that may require replacement of the piston. Examine the piston rings for the same conditions.

FIGURE 16.7 Location of the governor lever that must be removed. (*Lawn Boy Division, Outboard Marine Corporation.*)

FIGURE 16.6 Removing the governor yoke. (*Lawn Boy Division, Outboard Marine Corporation.*)

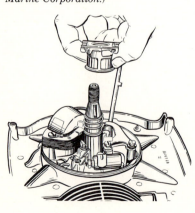

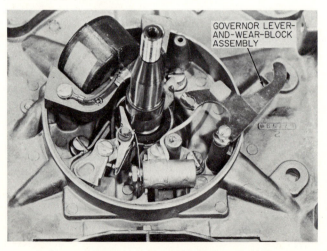

GOVERNOR LEVER-AND-WEAR-BLOCK ASSEMBLY

FIGURE 16.8 Removing screws so that the magneto plate can be removed. (*Lawn Boy Division, Outboard Marine Corporation.*)

FIGURE 16.9 Removing the connecting-rod-cap screws. (*Lawn Boy Division, Outboard Marine Corporation.*)

FIGURE 16.10 Detaching the cylinder head by removing the head bolts. (*Lawn Boy Division, Outboard Marine Corporation.*)

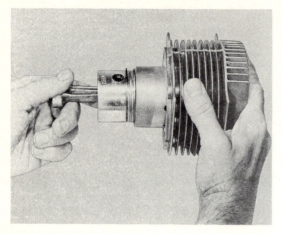

FIGURE 16.11 Taking piston-and-rod assembly from cylinder. (*Lawn Boy Division, Outboard Marine Corporation.*)

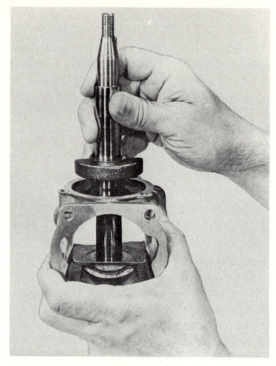

FIGURE 16.12 Removing the crankshaft from the crankcase. (*Lawn Boy Division, Outboard Marine Corporation.*)

If the cylinder wall shows signs of wear so that it will have to be refinished, discard the piston and rings because a new, oversized piston will be required to fit the enlarged cylinder. Cylinder service is discussed below.

Continue disassembly by removing the crankshaft from the crankcase through the top, as shown in Fig. 16.12. You have now disassembled the engine as shown in Fig. 16.13.

16.3 Servicing the Cylinder. Examine the cylinder for cracks, stripped threads in the bolt holes, broken fins, and scores or other types of damage in the cylinder bore. Any of these requires replacement of the cylinder. Quite often, however, stripped threads can be repaired with Heli-Coils. The repair is made by drilling out the worn

FIGURE 16.13 Major components of the engine. (*Lawn Boy Division, Outboard Marine Corporation.*)

threads, tapping the hole with a special Heli-Coil tap, and installing a Heli-Coil insert to bring the hole back to its original thread size, as shown in Fig. 16.14.

If the cylinder appears in good condition, use an inside micrometer or a dial indicator, as shown in Fig. 16.15, or else use a telescoping gauge and an outside micrometer, as shown in Fig. 16.16, to check the bore for taper or out-of-round wear.

If the cylinder bore is scored or worn, tapered or out of round, then it must be honed to a larger size so that a larger size piston and rings can be installed. Pistons are supplied in standard oversizes, for example, 0.010, 0.020, and 0.030 in. oversize, and the cylinder must be enlarged to take one of these oversize pistons.

Figure 16.17 shows the honing procedure to be used in the cylinder of a four-cycle one-cylinder engine. The procedure is similar for the cylinder of the two-cycle engine except it must be done from the crankcase end of the cylinder. Follow the instructions supplied by the manufacturer of the hone as to installing the hone on the cylinder, operating speed, and so on. Remove the hone and measure the cylinder periodically so you do not remove too much material. When the cylinder is approximately 0.002 in. within the desired size, change to fine stones

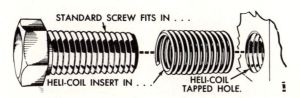

STANDARD SCREW FITS IN . . .

HELI-COIL INSERT IN . . .

HELI-COIL TAPPED HOLE.

FIGURE 16.14 Heli-Coil installation.

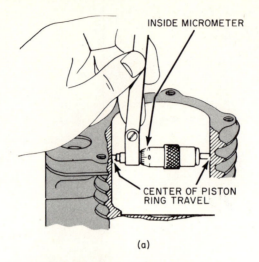

(a)

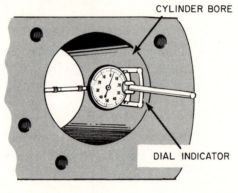

FIGURE 16.15 Methods of measuring cylinder bore. (*a*) Using an inside micrometer. (*b*) Using a dial indicator to check for irregular wear.

(b)

to finish the honing operation. When the honing job is finished, the cylinder wall should have a crosshatch pattern, such as shown in Fig. 16.18. This finish requires that the hone be moved up and down at the right speed while the hone is rotating at the right speed.

CAUTION. **The cylinder must be cleaned very carefully after the honing job is done, in order to remove all particles of grit and metal that may have become embedded in the cylinder wall. The best way to do this is to use soap and water and clean rags or a mop (Fig. 16.19). Wash the cylinder wall thoroughly or until you can rub a clean cloth on it without getting the cloth dirty. Then dry the cylinder wall and coat it with engine oil. Do not use kerosene or gasoline to clean the cylinder wall; they will not remove all the grit.**

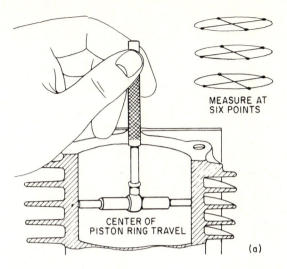

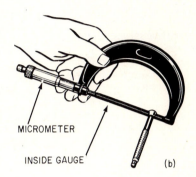

MEASURE AT
SIX POINTS

CENTER OF
PISTON RING TRAVEL

(a)

MICROMETER

INSIDE GAUGE

(b)

FIGURE 16.16 Taking bore measurements with a telescoping gauge and an outside micrometer. First you set the gauge to the bore diameter (*a*), and then you determine the diameter by using the micrometer as shown at (*b*).

16.4 Servicing Piston and Rings.

Rods are attached to pistons in several ways. In one, the piston pin is a press fit in the rod and must be driven out with a special punch. See Fig. 16.20. Another design

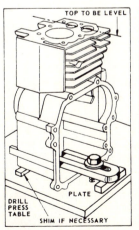

TOP TO BE LEVEL

DRILL
PRESS
TABLE

PLATE

SHIM IF NECESSARY

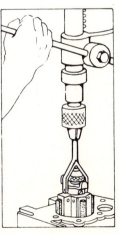

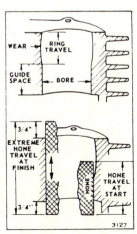

WEAR

RING
TRAVEL

GUIDE
SPACE

BORE

3/4"

EXTREME
HONE
TRAVEL
AT
FINISH

HONE

HONE
TRAVEL
AT
START

3/4"

3127

FIGURE 16.17 Honing the cylinder. (*Briggs and Stratton Corporation.*)

FIGURE 16.18 Cross-hatch appearance of a properly honed cylinder bore. (*Briggs and Stratton Corporation.*)

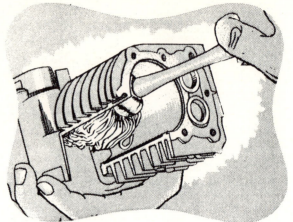

FIGURE 16.19 Cleaning the cylinder bore with a hand mop and soap and water. The cylinder has been cut away so the washing action can be seen. (*Clinton Engines Corporation.*)

uses retaining rings (also called lock rings) to hold the piston pin in place, as shown in Fig. 16.21. On these, needle-nose pliers must be used to remove the clips so the piston pin can be slipped out, as shown in Fig. 16.22. With the piston pin removed, the rod and piston may be separated. Some connecting rods are of the offset type, as shown in Fig. 16.23.

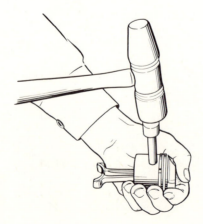

FIGURE 16.20 Removing the piston pin with a special punch. (*Lawn Boy Division, Outboard Marine Corporation.*)

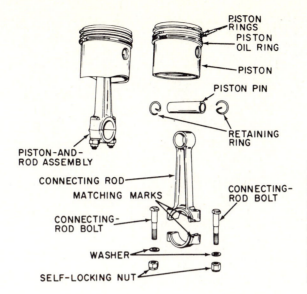

FIGURE 16.21 Piston, piston pin, and connecting rod, assembled to the left, and disassembled to the right. This design uses retaining or lock rings to hold the piston pin in position in the piston. (*Tecumseh Products Company.*)

Remove the piston rings, one at a time, using a ring remover, as shown in Fig. 16.24, to spread the rings so they can be slipped off over the head of the piston. If the rings appear to be worn or scuffed, discard them and install a new set of rings.

FIGURE 16.22 Using needle-nose pliers to remove lock rings. (*Briggs and Stratton Corporation.*)

FIGURE 16.23 Offset-type connecting rod. Some models have an oil dipper, as shown. (*Tecumseh Products Company.*)

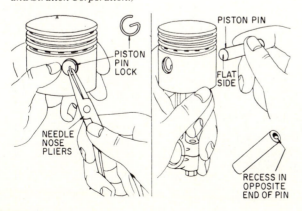

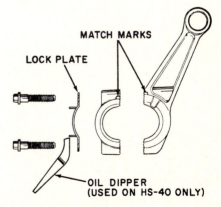

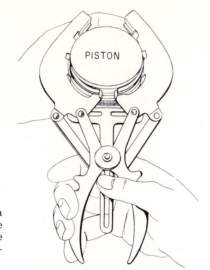

FIGURE 16.24 Using a ring expander to remove the piston rings from the piston. (*Briggs and Stratton Corporation.*)

Discard the piston if it is scored, shows wear spots, or is collapsed to an out-of-round condition. A piston should be checked with a micrometer, as shown in Fig. 16.25, and the dimensions compared to the cylinder dimensions (see Figs. 16.15 and 16.16). Some engine manufacturers recommend that the fit of the piston to the cylinder bore be checked by inserting the piston into the bore with a feeler gauge along its side. This will tell you whether or not there is too much clearance. Excessive wear will cause this condition. The remedy for excessive wear is to refinish the cylinder and install a new oversize piston. Always discard the old piston if the cylinder is to be refinished because if this happens a new, oversize piston must be installed.

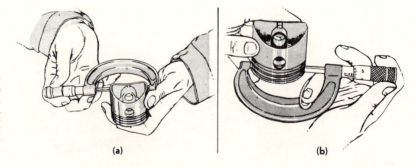

FIGURE 16.25 Measuring the diameter of the piston with outside micrometer. (*a*) Measuring skirt diameter. (*b*) Measuring piston-head diameter.

(a)   (b)

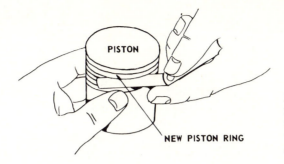

FIGURE 16.26 Check-
ing the side clearance
of the piston-ring groove.
(*Briggs and Stratton Corporation.*)

If the piston appears in good condition and is to be reused, check
for ring-groove wear by cleaning all carbon from the ring grooves. Put
a new ring of the proper size and kind in the top ring groove. Then
check the clearance between the ring and the side of the groove with
a feeler gauge, as shown in Fig. 16.26. If the clearance is excessive
(see manufacturer's specifications), discard the piston and install a
new one.

To check rings, clean all carbon from them, especially their ends,
and, checking one at a time, insert the ring in the cylinder bore,
pushing it down one inch into the bore. Square the ring with the
cylinder by pushing it down with a piston from which the rings have
been removed, as shown in Fig. 16.27. Check the ring gap with a feeler
gauge, as shown in Fig. 16.28. If it is excessive, throw the ring away.

Before installing the rings on the piston, check the fit of the piston
pin in the piston-pin bushings. You can measure the pin with an out-

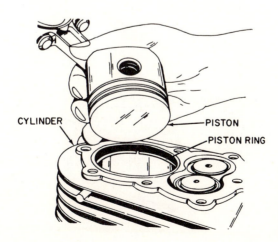

FIGURE 16.27 Squaring
ring in cylinder bore with
a piston in preparation for
measuring ring gap. (*Te-
cumseh Products Com-
pany.*)

FIGURE 16.28 Checking ring gap with a feeler gauge. (*Tecumseh Products Company.*)

side micrometer and the bushings with an inside gage and micrometer and compare the measurements. If they are not within the specifications in the manufacturer's service manual, correction must be made. Some manufacturers supply oversize piston pins. If these are available, you can bore the piston bushings and install the oversize pin. This means boring the connecting-rod bushing also. Other manufacturers supply new bushings which can be installed so that a standard-sized pin can be used. All this takes special equipment and is not recommended for a shop that does not have the proper tools.

To install rings on the piston, use the ring expander, as shown in Fig. 16.24. Stagger the ring gaps so they are evenly spaced around the piston. That is, with two rings, set the gaps 180° apart. With three rings, the gaps should be 120° apart. Figure 16.29 shows a set of rings installed on a piston. Note that the rings, on this piston, are of three

FIGURE 16.29 Positions of rings on piston. (*Briggs and Stratton Corporation.*)

types, and that the oil ring goes on the bottom. The ring gaps of the two upper rings are shown close together, but they should be 120° apart on this piston.

After the connecting rod has been reattached to the piston-and-ring assembly, the rings must be compressed into the piston-ring grooves in the piston before the assembly can be inserted into the cylinder bore. A ring compressor is required for this job. Figure 16.30 shows how the ring compressor is used. Cover the rings and cylinder with oil before installing the piston-ring-and-connecting rod assembly. This will allow the rings to slip into the cylinder more easily.

Note that, in the right-hand picture in Fig. 16.30, the piston is being installed in a four-cycle engine. The principle is the same whether it is a two-cycle or a four-cycle engine being serviced. There is this difference, however. In the two-cycle engine, the piston is pushed up into the cylinder from the crankcase side.

16.5 Connecting-rod Service. When the connecting rod is separated from the piston, check it for straightness and for cracks. The rod small end has a bushing in most engines. The bushing should be checked for wear, and if worn, should be replaced. On some models, the bushings are not separately replaceable. If the bushing is worn, the complete connecting rod must be replaced. And remember that the piston pin must properly fit both the connecting-rod bushing and the piston bushings, as we explained in a previous paragraph.

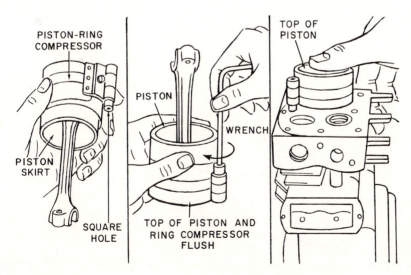

FIGURE 16.30 Using the ring compressor to install the piston-and-ring assembly in the cylinder bore. (*Briggs and Stratton Corporation.*)

Before you check the fit of the big-end bearing to the crankpin on the crankshaft, inspect the crankpin for wear and roughness, as explained in a following section on crankshaft service.

Some connecting rods have sleeve bearings in the big end; others use needle bearings. On the sleeve type, check the fit of the bearing to the crankpin on the crankshaft with shim stock or plastigage. If you use shim stock, put a 0.002-in. shim in the rod cap and install the rod and cap on the crank. Tighten the rod nuts and note whether or not the rod tightens up on the crankpin. If it does, then the clearance between the bearing and crank is less than 0.002 in. If it does not, put another shim on top of the first and try again. If it still does not tighten up, try an additional shim. In this way, you will find what the actual clearance is. Compare this with the manufacturer's specifications. Excessive clearance requires new bearings.

To use plastigage, lay a strip of plastigage on the bearing in the rod cap, as shown in Fig. 16.31. Then install the cap and rod on the crank and tighten the rod nuts. Remove the cap and measure the amount that the plastigage has flattened, as shown in Fig. 16.31. If the clearance is small, the plastigage will flatten considerably. If the clearance is large, the plastigage will flatten relatively little. Excessive clearance requires new bearings.

On some engines, the bearings are replaceable separately. On others, the complete rod must be replaced because the bearing is permanently bonded to the rod.

If the connecting-rod big end uses needle bearings, and the bearings are worn, they can be replaced. Figure 16.32 shows a connecting rod using needle bearings. Figure 16.33 shows two styles of needle bearings, split rows and single rows. New sets come in strips, as shown in Fig. 16.34. To install a new set of needles, lay the strip on the forefinger, as shown, and carefully strip off the backing. Then curl the finger around the crankpin so the needles transfer from your finger to the crankpin. The grease on the needles will hold them in place.

FIGURE 16.31 Using plastigage to check bearing clearance. (*a*) Lay a strip of plastigage on the bearing in the rod cap. (*b*) After installing the cap and then removing it, measure the amount the plastigage has flattened to determine clearance.

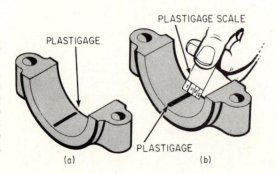

(a)   (b)

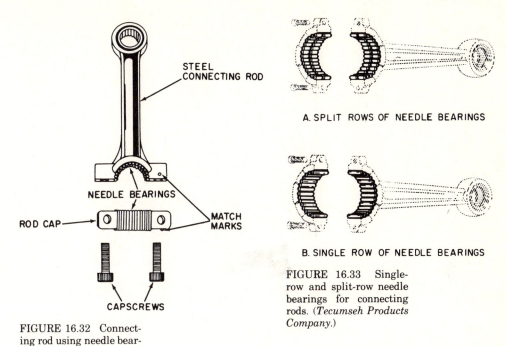

STEEL
CONNECTING ROD

NEEDLE BEARINGS

ROD CAP

MATCH
MARKS

CAPSCREWS

A. SPLIT ROWS OF NEEDLE BEARINGS

B. SINGLE ROW OF NEEDLE BEARINGS

FIGURE 16.32 Connect-
ing rod using needle bearings. (*Tecumseh Products
Company.*)

FIGURE 16.33 Single-
row and split-row needle
bearings for connecting
rods. (*Tecumseh Products
Company.*)

16.6 Crankshaft Service. Inspect the crankshaft crankpin and
main-bearing journals for signs of roughness. Figure 16.35 shows
various checks to be made on a crankshaft. The crankpin and journals
should be checked with a micrometer, as shown in Fig. 16.36. Checks
should be made all the way around each journal and the crankpin, and
also from one end to the other. If a journal or the crankpin is found to
be out of round more then 0.001 in., or if it is tapered, the crankshaft
should be discarded. It is possible to grind the journals and crankpin
to smaller diameters and use undersized bearings, but this is a delicate
operation that requires special equipment.

Battered threads on the crankshaft can be cleaned up with a
special thread die, called a thread chaser.

Sleeve-type crankshaft, or main, bearings should be checked for
wear by using plastigage, as already described in Sec. 16.5.

16.7 Gaskets. Always use new gaskets on engine reassembly. Old
gaskets are probably hard and would not provide a good seal. Make
sure that the sealing surfaces on the engine are clean, but do not

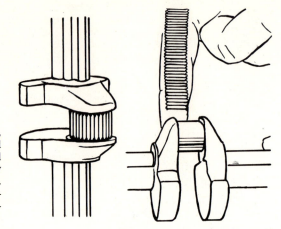

FIGURE 16.34 Left, needles in place around the crankpin. Right, hold needles on finger to apply them to the crankpin. (*Lawn Boy Division, Outboard Marine Corporation.*)

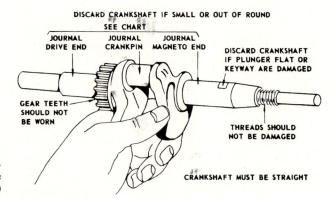

DISCARD CRANKSHAFT IF SMALL OR OUT OF ROUND
SEE CHART

JOURNAL DRIVE END   JOURNAL CRANKPIN   JOURNAL MAGNETO END

DISCARD CRANKSHAFT IF PLUNGER FLAT OR KEYWAY ARE DAMAGED

GEAR TEETH SHOULD NOT BE WORN

THREADS SHOULD NOT BE DAMAGED

CRANKSHAFT MUST BE STRAIGHT

FIGURE 16.35 Crankshaft check points. (*Briggs and Stratton Corporation.*)

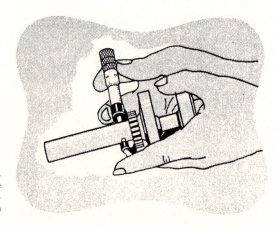

FIGURE 16.36 Using a micrometer to measure crankshaft journal. (*Clinton Engines Corporation.*)

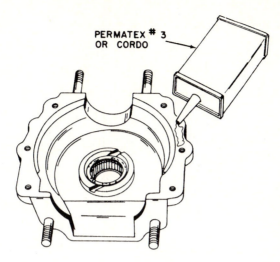

PERMATEX # 3
OR CORDO

FIGURE 16.37 Applying sealant to the contact face of one half of a split crankcase. (*Tecumseh Products Company.*)

scrape them. Instead, to remove traces of sealer or gasket material, use lacquer thinner and a clean cloth to wipe the surfaces.

On the split-crankcase engine, the two halves of the crankcase are sealed by a bead of liquid gasket cement applied to the contact face of one of the halves, as shown in Fig. 16.37.

16.8 Reed Valves. Make sure that the reeds are not bent or damaged and that they seal tightly when they close. Replace the reed-valve assembly if the reeds are not in good condition.

> CAUTION. **Do not attempt to check the reeds with compressed air; you could damage them.**

16.9 Ball and Roller Bearings. If ball bearings are in good condition, leave them in the crankcase or on the crankshaft. You can check them by rotating the inner race to see whether they turn roughly or hard. If the bearings are worn, they must be replaced. If the bearings are on the crankshaft, they should be pulled with a puller, as shown in Fig. 16.38, or pressed off with an arbor press, as shown in Fig. 16.39. Then a new bearing can be pressed on with a special tool.

If the bearings are in the crankcase, they may be pressed out or knocked out with a special tool, as shown in Fig. 16.40. Another removal method, recommended by some engine manufacturers, is to put the crankcase half on a hot plate, as shown in Fig. 16.41. As the crankcase half reaches a temperature of about 400°, the ball bearing should drop out. You might have to tap the case lightly with a soft

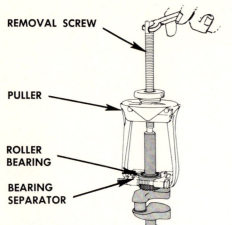

FIGURE 16.38 Removing ball bearing from crankshaft with a bearing puller.

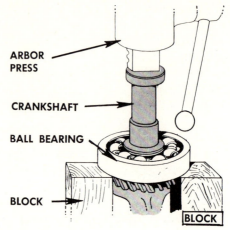

FIGURE 16.39 Pressing ball bearing off crankshaft in an arbor press.

FIGURE 16.40 Using special tool to remove or replace ball bearing in crankcase section. (*Lawn Boy Division, Outboard Marine Corporation.*)

FIGURE 16.41 Using a hot plate to heat the crankcase half and make it easy to remove ball bearing. (*Tecumseh Products Company.*)

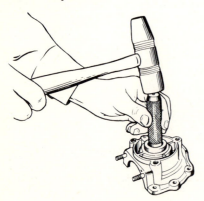

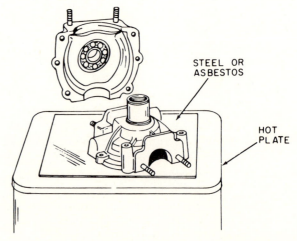

hammer to assist in the loosening process. Then the new bearing can be dropped into the case. Make sure it goes all the way into the recess so it is seated. You will need heavy gloves to handle the hot case.

If the crankshaft is mounted on roller bearings, as shown in Fig. 16.42, you can replace the outer race in the housing by pulling it or heating the housing and then installing a new race. The inner race on the crankshaft must be pulled with a special puller and a new race pressed on.

16.10 Oil Seals.   Oil seals must be discarded and new seals installed every time the engine is torn down. One oil-seal arrangement is shown in Fig. 16.43. The seal is held in place by a retainer and a snap ring, as shown. In order to remove the snap ring and the seal, you use an ice pick or other pointed tool to pry the snap ring out of the spring groove. This permits removal of the spring, retainer, and seal. Figure 16.44 shows the point at which the tool must be inserted to remove the snap spring.

FOUR-CYCLE ENGINE SERVICE

16.11 General Instructions.   Many of the services previously described for two-cycle engines also apply to four-cycle engines. However, there is one service that a four-cycle engine has all to itself, and this concerns valves. As you know, the four-cycle engine must have two valves per cylinder, an intake valve and an exhaust valve. In this part of the chapter, we will describe the services that engines with valves—four-cycle engines—requires insofar as the valves are concerned. This includes removing the cylinder head, removing and

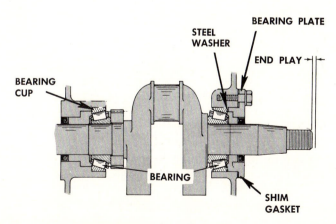

FIGURE 16.42 Crankshaft mounted on tapered roller bearings. (*Kohler Company.*)

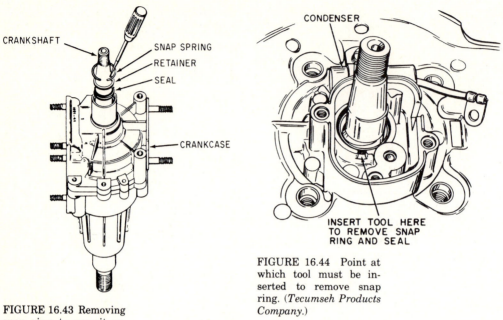

FIGURE 16.43 Removing snap ring to permit removal of oil seal. (*Tecumseh Products Company.*)

FIGURE 16.44 Point at which tool must be inserted to remove snap ring. (*Tecumseh Products Company.*)

servicing valves, servicing valve seats and guides, adjusting valve tappet clearance, installing new valve-seat inserts, and checking and servicing the camshaft and bearings as required.

NOTE: The procedures that follow apply generally to late-model Briggs and Stratton engines. They may also apply in a general way to engines made by other companies. However, to be on the safe side, always have the complete instructions in front of you whenever you start out to service a particular model of engine.

16.12 Removing and Replacing Cylinder Head. As you remove the cylinder-head screws, be sure to note where the long and short screws go. If you try to put a long screw in a short-screw hole, it might bottom and break off a fin or leave the cylinder head loose. If you put a short screw in a long-screw hole, it may not engage enough threads to hold and will either strip the upper threads or else not retain its hold so the cylinder head will be loose. See Fig. 16.45.

When replacing the cylinder head, use a new head gasket. Do not use any sealer on the gasket! Use graphite grease on screws that go into aluminum cylinders. Tighten the screws down evenly by hand and then use a torque wrench to finish the job. Tighten the screws in the sequence shown in Fig. 16.45 and to the torques indicated in the table in Fig. 16.46.

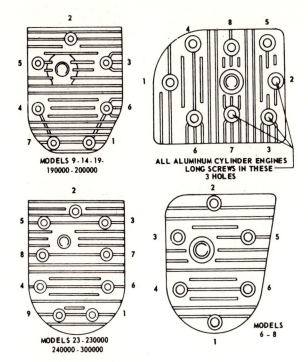

BASIC MODEL SERIES	IN.-LB TORQUE
ALUMINUM CYLINDER	
6B, 60000, 8B, 80000 82000, 92000 100000, 130000	140
140000, 170000	165
CAST IRON CYLINDER	
5, 6, N, 8, 9	140
14	165
19, 190000, 200000, 23 230000, 240000, 300000	190

FIGURE 16.46 Table showing proper torques for tightening cylinder head screws. (*Briggs and Stratton Corporation.*)

FIGURE 16.45 Cylinder heads for various small engines, showing torquing sequence, that is, the sequence in which the cylinder head screws should be tightened. (*Briggs and Stratton Corporation.*)

CAUTION. **Do not tighten the screws down to the proper tension the first time you put the wrench on them. Instead, tighten each screw in sequence only a little. Go around again and a third or fourth time, tightening the screws a little more each time until finally all are at the proper tightness. This assures even tension on all screws and guards against a warped cylinder head.**

16.13 Removing Valves.

There are three general types of valve-spring retainers, as shown in Fig. 16.47, pin, split-collar, and one-piece. To remove the pin or split type, use a spring compressor, as shown in Fig. 16.48 and 16.49. Adjust the jaws of the compressor until they just touch the top and bottom of the valve chamber. This will keep the upper jaw from slipping into the coil of the spring. Push the compressor in until the upper jaw slips over the upper end of the spring.

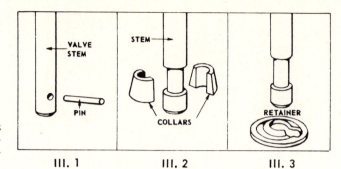

FIGURE 16.47 Different types of valve-spring retainers. (*Briggs and Stratton Corporation.*)

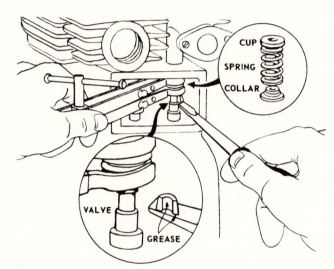

FIGURE 16.48 Removing spring on engine using split collar. (*Briggs and Stratton Corporation.*)

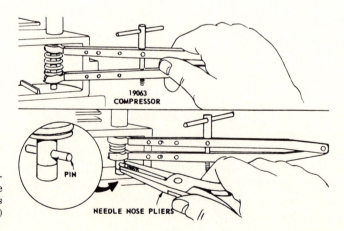

FIGURE 16.49 Spring-removal method on engine using pin retainer. (*Briggs and Stratton Corporation.*)

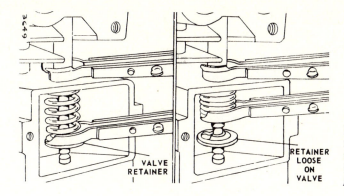

FIGURE 16.50 Spring-removal method on engine using one-piece retainer. (*Briggs and Stratton Corporation.*)

Then compress the spring by tightening the jaws. On the split-collar type, put a little grease on a screwdriver, as shown in Fig. 16.48, to remove the retainer. On the pin-type, use needle-nose pliers to pull out the pin. See Fig. 16.49.

On the one-piece type, Fig. 16.50, move the retainer around so the larger part of the opening clears the undercut in the valve. Then, on all types, lift the valve out and remove the compressor with the spring.

16.14 Servicing Valves. Figure 16.51 shows the various parts of the valve to be inspected. Check the valve seating faces for wear, burned spots, pits, cracks, and other signs of damage. If the face

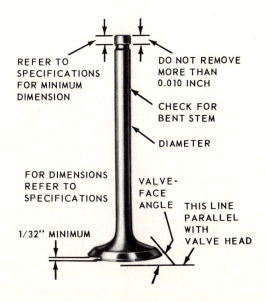

FIGURE 16.51 Valve parts to be checked. On the valve shown, the stem end is hardened and no more than about 0.010 in. should be removed. (*Ford Motor Company.*)

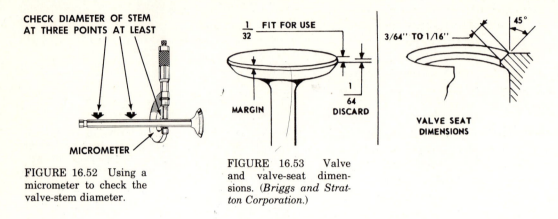

CHECK DIAMETER OF STEM AT THREE POINTS AT LEAST

MICROMETER

FIGURE 16.52 Using a micrometer to check the valve-stem diameter.

$\frac{1}{32}$ **FIT FOR USE**

MARGIN

$\frac{1}{64}$ **DISCARD**

FIGURE 16.53 Valve and valve-seat dimensions. (*Briggs and Stratton Corporation.*)

3/64" TO 1/16" 45°

VALVE SEAT DIMENSIONS

seems to be in good condition but is somewhat worn or burned, it can be refaced on a special machine. Also, check the valve stem with a micrometer, as shown in Fig. 16.52. Figure 16.53 shows typical valve and seat dimensions as recommended by Briggs and Stratton. If the valve face looks to be too badly worn or otherwise damaged to clean up in the valve-refacing machine, or if the valve stem is worn or bent, discard the valve.

Figure 16.54 shows a valve-refacing machine. It is used by clamping the valve stem in the chuck of the machine and bringing the

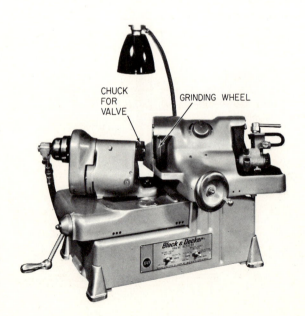

CHUCK FOR VALVE

GRINDING WHEEL

FIGURE 16.54 Valve-refacing machine, also called a valve grinder. (*Black and Decker Manufacturing Company.*)

seating face into contact with the rotating grinding wheel. As the valve face is rotated against the wheel, it is smoothed and all irregularities are removed. If cleaning up the face reduces the margin too much (see Fig. 16.53), discard the valve. A valve with an excessively small margin will overheat and soon burn. Refer to the operating instructions of the valve-refacing machine for details of how to operate it.

16.15 Checking Valve Springs and Tappets. Valve springs should be tested for proper tension and for squareness. A special fixture is required to check spring tension (Fig. 16.55). The pressure required to compress the spring to the proper length should be measured. Then the spring should be checked for squareness, as shown in Fig. 16.56. Stand the spring on a surface plate and hold a steel square next to it, as shown. Rotate the spring slowly against the square and see whether the top coil moves away from the square. If the spring is excessively out of square or lacks sufficient tension, discard it.

Check the valve-tappet faces which ride on the cams for roughness or wear. Check the adjusting-screw head which is in contact with the valve stem, or the stem end of the tappet on tappets which do not have adjusting screws, for wear. If you find wear or roughness, new tappets will be required. The camshaft must be removed to remove the tappets as explained in Sec. 16.19.

16.16 Servicing Valve Seats. If the valve seat is worn, burned, pitted, or otherwise damaged, it should be ground with a special seat grinder. If the seat is so badly worn or burned so that it will not clean

FIGURE 16.55   Checking spring tension with a special fixture. (*Ford Motor Company.*)

FIGURE 16.56   Checking spring squareness. (*Ford Motor Company.*)

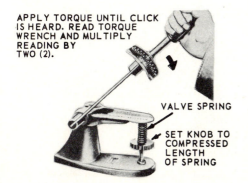

up, it is possible to counterbore the seat and install a seat insert, as we will explain in a following section. As a first step in grinding a valve seat, you should make sure that the valve guide is in good condition. The reason for this is that the seat grinder is centered in the valve guide. We will explain how to check and service valve guides in a following section.

Figure 16.57 shows how the seat grinding stone is centered by means of a pilot installed temporarily in the valve guide. In operation, an electric motor rotates the grinding stone and smooths the seat. The angle of the stone determines the angle to which the seat will be ground. This angle matches the angle to which the valve face is ground in the valve-face regrinder.

Many engine manufacturers recommend that after the valve and valve seat have been ground, the two should be lapped together to perfect the fit. This operation is shown in Fig. 16.58. The valve lapping tool has a suction cup which holds the valve head. Lapping compound (an abrasive paste) is applied to the valve face, and the valve is then rotated against the valve face. After the lapping operation is completed, the cylinder block and valve should be cleaned up to remove all traces of lapping compound.

FIGURE 16.57   Pilot on which the grinding stone rotates. The pilot keeps the stone concentric with the valve seat. (*Black and Decker Manufacturing Company.*)

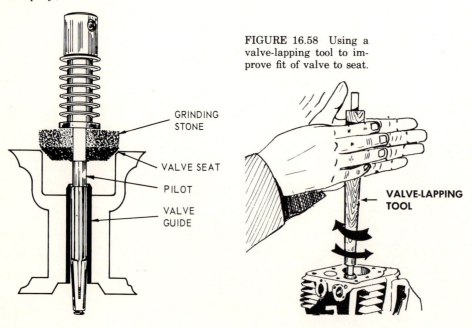

FIGURE 16.58   Using a valve-lapping tool to improve fit of valve to seat.

GRINDING STONE

VALVE SEAT

PILOT

VALVE GUIDE

VALVE-LAPPING TOOL

16.17 Valve-seat Inserts. On those engines with valve-seat inserts, the inserts must be replaced if they cannot be ground satisfactorily to restore good valve seating. If the cylinder block does not have a valve-seat insert fitted, an insert can be installed after reboring the block. Figures 16.59 and 16.60 show how to install the counterbore cutter, and Fig. 16.61 shows the counterboring procedure. If a valve seat must be replaced, the old seat must be pulled with a special tool, as shown in Figs. 16.62 and 16.63. Turning the bolt will pull the seat.

To install a new valve seat, the proper pilot must be used, along with the correct driver. Insert the pilot into the valve guide, and then drive the insert into place with the driver, as shown in Fig. 16.64. It should then be ground lightly and the valve lapped into the seat with lapping compound.

On aluminum blocks, peen around the insert, as shown in Fig. 16.65.

NOTE: The valve guide must be in good condition and without excessive wear. If it is worn, the counterbore might not be centered.

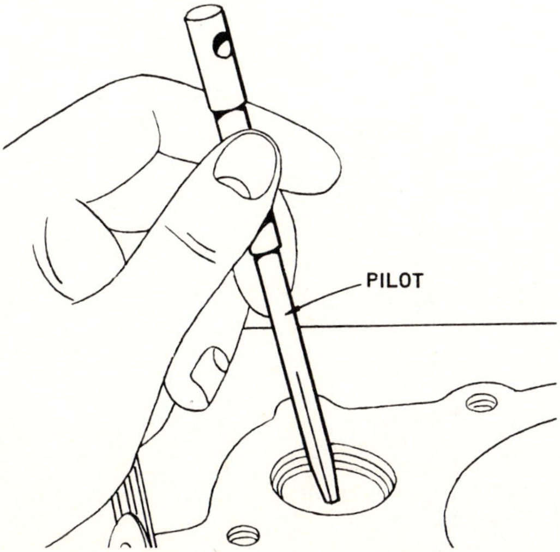

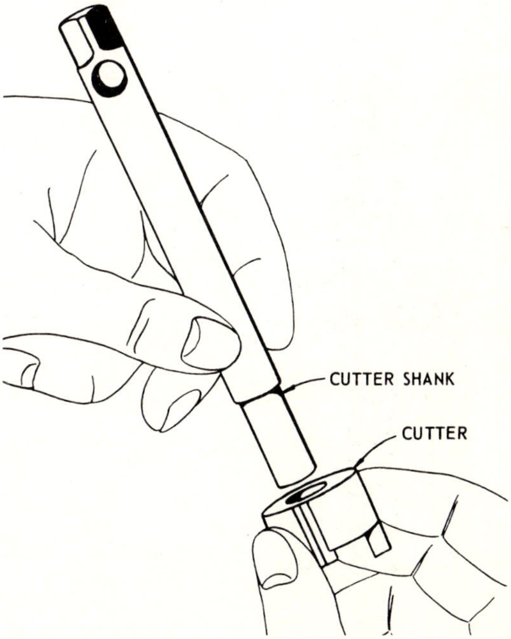

FIGURE 16.60 Inserting cutter shank in cutter. (*Briggs and Stratton Corporation.*)

FIGURE 16.59 Inserting pilot in preparation for counterboring for a valve-seat insert. (*Briggs and Stratton Corporation.*)

PILOT

CUTTER SHANK

CUTTER

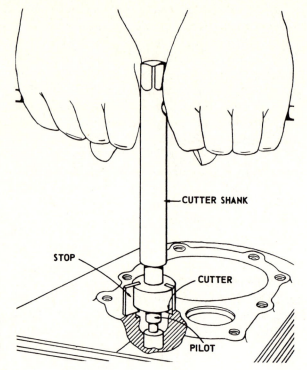

FIGURE 16.61 Counter-
boring for valve-seat in-
sert. (*Briggs and Stratton
Corporation.*)

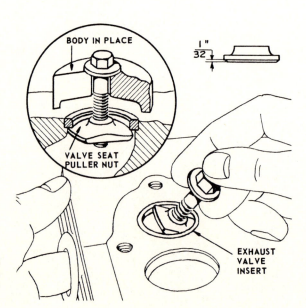

FIGURE 16.62 Inserting
valve seat-insert puller.
(*Briggs and Stratton Corporation.*)

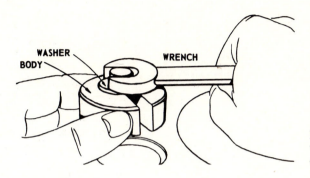

FIGURE 16.63 Pulling valve-seat insert. (*Briggs and Stratton Corporation.*)

16.18 Valve Guides. To check for valve-guide wear, use a plug gauge, as shown in Fig. 16.66. If it is worn, it should be rebushed, as shown in Fig. 16.66. First, a bushing guide is used to center the reamer and then the reamer is turned to ream out the worn guide. Ream to

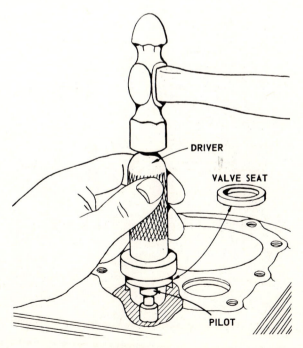

FIGURE 16.64 Driving in a new valve-seat insert. (*Briggs and Stratton Corporation.*)

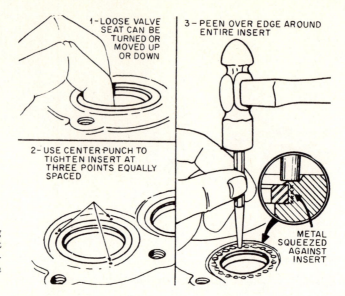

FIGURE 16.65 Peening around edge of valve-seat insert to hold it in position. (*Briggs and Stratton Corporation.*)

only $\frac{1}{16}$ in. deeper than the valve-guide bushing. Then, after cleaning out all cuttings, press in a new valve-guide bushing, using a soft metal driver of brass or copper to avoid peening over the top end of the bushing. Finally, finish reaming the new bushing to the proper size so that a standard valve can be used.

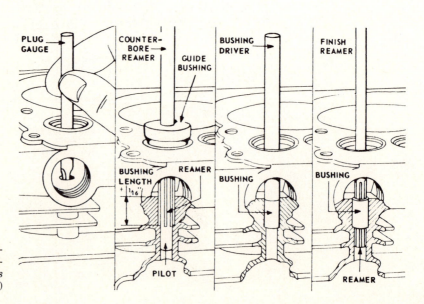

FIGURE 16.66 Checking, installing, and reaming valve guides. (*Briggs and Stratton Corporation.*)

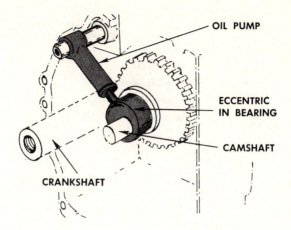

OIL PUMP

ECCENTRIC
IN BEARING

CAMSHAFT

CRANKSHAFT

FIGURE 16.67 Oil pump
driven by an eccentric on
the camshaft. (*Tecumseh
Products Company.*)

16.19 Camshaft Service. In addition to having a pair of cams to
operate the valves, camshafts may also have an eccentric to operate
the oil pump, as shown in Fig. 16.67. The action of this pump is shown
in Figs. 6.4 to 6.6. Some camshafts also include an automatic compression release device which is discussed in Sec. 4.8. See Fig. 4.12. In
addition, some earlier Kohler models had the ignition breaker-point
cam on the camshaft mounted through a centrifugal advance mechanism. This provided an ignition retard during cranking so as to avoid
kickback. After the engine started, the ignition would advance to the
running position. Later models, using the automatic compression
release, do not use this ignition advance mechanism. Also, some
models have a power takeoff on the camshaft, as shown in Fig. 16.68.
Some models also have a governor drive gear which drives an internal

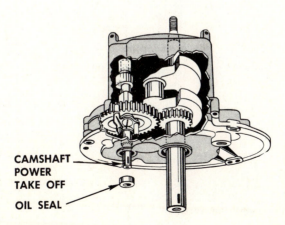

CAMSHAFT
POWER
TAKE OFF

OIL SEAL

FIGURE 16.68 Engine
partly cut away so the
crankshaft and camshaft
with power takeoff can be
seen. (*Wisconsin Motor
Corporation.*)

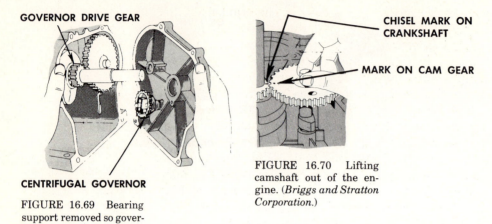

GOVERNOR DRIVE GEAR

CENTRIFUGAL GOVERNOR

FIGURE 16.69 Bearing support removed so governor drive gear can be seen.

CHISEL MARK ON CRANKSHAFT

MARK ON CAM GEAR

FIGURE 16.70 Lifting camshaft out of the engine. (*Briggs and Stratton Corporation.*)

centrigugal governor, as shown in Fig. 16.69. So you see that you are apt to meet up with a variety of camshafts on four-cycle engines.

If the camshaft or bearings appear to be worn, then you will need to remove the camshaft for inspection. The removal procedure varies according to the engine design. On many models, you take off the baseplate or gear cover and lift out the camshaft (Fig. 16.70). Be sure to find the timing marks on the camshaft and crankshaft gears. These can be center-punch or chisel marks or, on some models, a chamfer on the end of one crankshaft-gear tooth (Fig. 16.71). These marks must be aligned when the camshaft is reinstalled to assure proper valve timing.

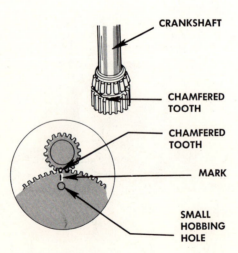

CRANKSHAFT

CHAMFERED TOOTH

CHAMFERED TOOTH

MARK

SMALL HOBBING HOLE

FIGURE 16.71 Timing marks on crankshaft gear (a chamfered tooth), and on the camshaft gear (a chisel mark). (*Tecumseh Products Company.*)

On some engines, you will have to remove the crankshaft first before you can remove the camshaft. Also, one model has an end-thrust bearing which must be removed when you remove the camshaft.

Before attempting to remove the camshaft, turn the camshaft until the timing marks align on the compression stroke. This means that both valves are closed and that there is no pressure on the valve tappets. This makes it easy to slip the camshaft out. Turn the engine on its side so the tappets will not fall out. Note that some engine models using ball bearings require removal of the camshaft and crankshaft together. Note also that some models have the camshaft supported on a pin which must be driven out before the camshaft can be removed. If the camshaft has an end-play washer, remove it and lay it aside. It should be put back in the same position that you found it.

Figure 16.72 shows the various points at which the camshaft should be inspected. Wash the camshaft with solvent to clean off dirt and oil. Check the gears for wear or nicks. Check the automatic compression release or ignition advance mechanism if present, for freeness of action. If the camshaft has oil holes, blow them out with compressed air. The camshaft dimensions, arrowed in Fig. 16.72, should then be checked with a micrometer and the measurements compared with the specifications in the repair manual for the engine you are working on. Discard the camshaft if the wear is excessive.

Remove the tappets and examine them as already explained in Sec. 16.15.

If the camshaft rides in sleeve bearings, or bushings, check them for wear and replace them if necessary. Replacement requires special

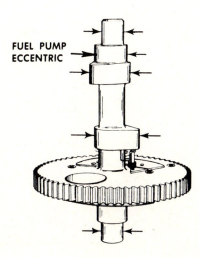

FUEL PUMP
ECCENTRIC

FIGURE 16.72 Points at which the camshaft should be checked. (*Tecumseh Products Company.*)

drivers to force the old bushings out and new bushings in. Then, the new bushings must be reamed to size.

If the camshaft is supported by ball or roller bearings, check them as already explained in Sec. 16.9.

To replace the camshaft, first install the tappets. Make sure you put the tappets back in the same holes from which you took them. If you get the tappets reversed, you are probably in for trouble because they may not fit properly. On some models, the two tappets are of different lengths.

Push the tappets up out of the way and then install the camshaft. If the camshaft is of the type supported by a camshaft pin, put the camshaft and crankshaft in first and align the timing marks before positioning the camshaft and driving the camshaft pin into place.

On all models, be sure the timing marks are properly aligned when installing the camshaft and crankshaft. If the camshaft has additional parts attached, such as the oil pump, governor drive gear, ignition centrifugal advance mechanism, or automatic compression release, make sure they are properly aligned. Put a little oil on all parts—tappets, bushings, camshaft, advance or compression release —before final installation.

16.20 Installing Valves. Be sure to check the springs carefully, as already noted, before installing the valves. If the spring is held in place by a pin or a split collar, put the spring with the retainer in the spring compressor, as shown in Fig. 16.73. Insert the compressor with spring and retainer into position in the cylinder block and then drop the valve into place. Install the retainer pin or collar and release the spring pressure. Pull out the compressor. On the one-piece retainer, move the retainer around when dropping the valve into place so the valve stem enters the larger part of the opening, allowing the stem

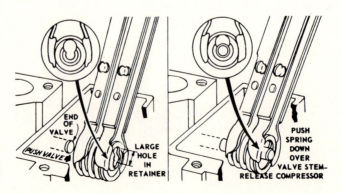

FIGURE 16.73 Compressing valve spring in preparation for valve installation. (*Briggs and Stratton Corporation.*)

to go through. Then lift the retainer up and center it in the undercut on the valve stem. Now release the spring pressure and remove the spring compressor.

16.21 Checking Valve-tappet Clearance. After valves, seats, and guides have been checked and serviced as necessary, install the valves in their proper positions in the cylinder block. Turn the crankshaft until one of the valves is in its highest position. Then turn the crankshaft one more complete revolution. Check the clearance with a feeler gauge. Repeat for the other valve. If the clearance is too small, as it may be if valves and seats have been ground, grind off the end of the valve stem, as necessary, to get the correct clearance.

Some valve tappets have an adjusting screw and, on these, you can adjust the valve-tappet clearance by turning the screw in or out, as shown in Fig. 16.74. On the model shown, the adjusting screw has a lock nut which must be loosened with one wrench while the second wrench holds the adjusting screw. Then the second wrench can be turned to make the adjustment. Finally, the lock nut is tightened to hold the adjusting screw in the correct position.

NOTE: Clearances on the small engine shown being serviced are checked with the engine cold. In automotive engines, the clearances are usually checked with the engine running and at operating temperature.

CHECKUP

In this final chapter in the book, you learned the fundamentals of small-engine service including disassembly, cylinder honing, checking and installing pistons, rings, connecting rods, and bearings, and how to service other engine components including crankshafts, valves, and oil seals. The following questions will not only give you a chance to check up on how well you understand and remember these fundamentals, but also will help you to remember them better. The act of writing down the answers to the questions will fix the facts more firmly in your mind.

FIGURE 16.74 Adjusting valve tappet, or valve lash, on an L-head engine.

NOTE: Write down your answers in your notebook. Then later you will find your notebook filled with valuable information which you can refer to quickly.

Completing the Sentences: Test 16. The sentences below are not complete. After each sentence there are several words or phrases, only one of which will correctly complete the sentence. Write each sentence in your notebook, selecting the proper word or phrase to complete it correctly.

1. The engine that manufacturers supply which is not quite a complete engine is called (*a*) a short-block engine; (*b*) a rebuilt engine; (*c*) an engine without piston or crankshaft.
2. Stripped threads in the block can often be repaired with (*a*) Permatex; (*b*) threaders; (*c*) a Heli-Coil insert.
3. After the honing operation, the best thing to use to clean the cylinder wall is (*a*) gasoline; (*b*) kerosene; (*c*) soap and water.
4. Ring gap should be checked with the ring (*a*) on the piston; (*b*) in the cylinder; (*c*) on the workbench.
5. When using plastigage, the bearing clearance is indicated by (*a*) how long the plastigage becomes; (*b*) flattening of the plastigage; (*c*) ease of rotating the crankshaft.
6. When grinding valve seats, the grinding stone is held in a pilot that is installed (*a*) on the cylinder head; (*b*) in the valve seat; (*c*) in the valve guide.
7. Valve-tappet clearance on some engines are adjusted by (*a*) grinding the valve stem; (*b*) grinding the valve; (*c*) replacing the tappet.
8. The three common types of valve-spring retainers are the pin, (*a*) pawl, and split; (*b*) split collar, and one-piece; (*c*) one-piece, and double-pin.

Written Checkup

In the following, you are asked to write down, in your notebook, the answers to the questions asked or to define certain terms. Writing the answers down will help you to remember them.

1. What is meant by a short-block engine?
2. Describe the procedure of disassembling a two-cycle engine.
3. Explain how to check a cylinder and service it if it is worn.
4. Explain how to check pistons and piston rings for wear and fit.
5. Explain how to install piston rings and piston in the engine.
6. Explain how to install needle bearings.

7. Explain how to use plastigage to determine sleeve-bearing clearance.
8. Explain how to check a crankshaft.
9. Describe the procedure of removing and replacing the valves in a four-cycle engine.
10. Describe the procedure of checking and servicing valves.
11. Explain how to check valve springs.
12. Describe the procedure of servicing valve seats.
13. Explain how to install a valve-seat insert.
14. Describe the procedure for checking a valve quide for wear and for installing a valve guide.
15. Describe the procedure of installing valves and checking the tappet clearance.

ANSWERS TO COMPLETING-THE-SENTENCE TESTS IN SMALL ENGINES

	Question									
Test	1	2	3	4	5	6	7	8	9	10
1	No test in Chap. 1									
2	(c)	(b)	(c)	(a)	(b)	(a)	(c)	(b)	(a)	(a)
3	(b)	(b)	(c)	(b)	(c)	(a)	(a)	(c)	(b)	(c)
4	(a)	(a)	(c)	(b)	(a)	(c)	(b)	(a)	(b)	(c)
5	(a)	(c)	(a)	(b)	(b)	(a)	(c)	(b)	(c)	(c)
6	(b)	(c)	(c)	(a)	(c)	(c)	(a)	(c)		
7	(c)	(b)	(b)	(a)	(b)	(c)	(a)	(b)	(a)	(b)
8	(c)	(b)	(c)	(a)	(b)	(c)	(a)	(c)	(b)	(b)
9	(c)	(b)	(a)	(c)	(a)	(c)	(b)	(c)	(a)	(c)
10	(b)	(b)	(c)	(c)	(a)	(c)	(b)	(a)		
11	(b)	(a)	(c)	(c)	(a)	(c)	(c)	(c)		
12	(a)	(c)	(a)	(b)	(a)					
13	(c)	(b)	(c)	(a)	(c)	(a)	(b)	(c)	(c)	(a)
14	(b)	(c)	(b)	(b)	(c)	(a)	(b)	(a)		
15	(b)	(c)	(b)	(a)	(b)	(b)	(c)	(a)		
16	(a)	(c)	(c)	(b)	(b)	(c)	(a)	(b)		

Index